The Invention of the Steam Engine

B. J .G. van der Kooij

This case study is part of the research work in preparation for a doctorate-dissertation to be obtained from the University of Technology, Delft, The Netherlands (www.tudelft.nl). It is one of a series of case studies about "Innovation" under the title "The Invention Series."

About the text: This is a scholarly case study describing the historic developments that resulted in the steam engine. It is based on a large number of historic and contemporary sources. As we did not conduct any research into primary sources, we made use of the efforts of numerous others by citing them quite extensively to preserve the original character of their contributions. Where possible we identified the individual authors of the citations. As some are not identifiable, we identified the source of the text. Facts that are considered to be of a general character in the public domain are not cited.

About the pictures: Many of the pictures used in this case study were found at websites accessed through the Internet. Where possible they were traced to their origins, which, when found, were indicated as the source. As most are out of copyright, we feel that the fair use we make of the pictures to illustrate the scholarly case is not an infringement of any possible copyright that sometimes (still) is claimed.

Copyright © 2015 B. J. G. van der Kooij

Cover art is line drawing of Savery's "Miner's Friend" (Wikimedia Commons)

Version 1.1 (January 2015)

All rights reserved.

ISBN-10: 1502809095
ISBN-13: 978-1502809094

Contents

Contents iii
Preface v
Context for the Discoveries 3
   The seventeenth and eighteenth centuries 4
      Turmoil in Europe 4
   England in the seventeenth and eighteenth centuries 7
      The Spirit of Time 7
      Living and working 9
      Land and water transportation 11
      Poor plebs and rich gentry 12
      Mining and water 13
   Science discovers and applies steam 19
      Steam as a phenomenon: power of fire 19
      Steam explored by engineering 26
   Applying the power of fire 33
      Savery's "Miner's Friend" (1698) 33
      Improvement of Savery's pump 36
      The power of fire understood 37
First-Generation Steam Engines (1700-1775) 39
   Aristocracy: the gentlemen of science 39
      Robert Boyle 40
      European influences 41
   Atmospheric engines 42
      Newcomen's atmospheric steam engine (1712) 43
      Improvements of the Newcomen steam engine 48
      A cluster of innovations 49

Second-Generation steam engines (1775-1800) 51
The discovery of external cooling 53
- James Watt's steam engine (1769) 54
- James Watt and his partners in business 57
- Perfecting the engine 62

Contemporary developments 65
- Bull's steam engine 66
- Hornblower's compound steam engine 66

Patent war 68
- Boulton & Watt: the end of a partnership 72

Applications of the steam engine 73
- William Symington's steam engine 73
- A cluster of innovations 75

Third-Generation Steam Engines (1800+) 77
High pressure eliminates the external condenser 77
- Trevethick's steam engine (c. 1802) 80
- Trevithick's other activities 86
- A cluster of innovations 91

Contemporary developments 92
- Steam engines built in the eighteenth century 92
- Stationary applications for steam engines 93
- Mobile applications for steam engines 94
- Rainhill Trials 103

Mobility infrastructures 106
- From wagonway to railroad 107
- From freight to passengers 108

Conclusion 109
References 113
About the Author 117

Preface

When everything is said and done,
and all our breath is gone.
The only thing that stays,
Is history, to guide our future ways.

My lifelong intellectual fascination with technical innovation within the context of society started in Delft, the Netherlands, in the 1970s at the University of Technology, both the Electrical Engineering School and the Business School[1]. Having been educated as a technical student with vacuum tubes, followed by transistors, I found the change and novelty caused by the new technology of microelectronics to be mind-boggling, not so much from a technical point of view but with all those opportunities for new products, new markets, and new organizations, with a potent technology as the driving force.

During my studies at both the School of Electric Engineering and the School of Business Administration,[2] I was lucky enough to spend some time in Japan and California, noticing how cultures influence the context for technical-induced change and novelty. In Japan I touched on the research environment; in the Silicon Valley, it was the business environment—from the nuances of the human interaction of the Japanese, to the stimulating and raw capitalism of America. The technology forecast of my engineering thesis made the coming technology push a little clearer: the personal computer was on the horizon. The implementation of innovation in small and medium enterprises, the subject of my management thesis, left a lot to question. Could something like a Digital Delta be created in the Netherlands?

---

[1] In the present time it is the Electrical Engineering School at the Delft University of Technology, and the School of International Business Administration at the Erasmus University Rotterdam.

[2] The actual names were Afdeling Electro-techniek, Vakgroep Mikro-Electronica, and Interfaculteit Bedrijfskunde.

During the journey of my life, innovation was the theme. For example, on the level of the firm, when in the mid-1970s I joined a mature electric company manufacturing electric motors, transformers, and switching equipment, business development was a major responsibility. How could we change an aging corporation by picking up new business opportunities? Japan and California were again on the agenda, but now from a business point of view: acquisition, cooperation, and subcontracting. Could we create business activity in personal computers?
The answer was no.

Innovation on the national level became the theme as I entered politics (a quite innovative move for an engineer) and became a member of the Dutch Parliament. How could we prepare a society for the new challenges that were coming, threating the existing industrial base and creating new firms and industries? What innovation policies could be applied? Introducing in the early 1980s the first personal computer in Parliament made me known as "Mr. Innovation" within the small world of my fellow parliamentarians. Could we, as politicians, change Dutch society by picking up the new opportunities technology was offering?
The answer was no.

The next phase on my journey brought me in touch with two extremes. A professorship in the Management of Innovation at the University of Technology in Eindhoven gave room for my scholarly interests. I was (part-time) looking at innovation at the macro level of science. The starting of a venture company making application software for personal computers satisfied my entrepreneurial obsession. Now it was about the (nearly full-time) implementation of innovation on the microscale of a starting company. With both my head in the scientific clouds and my feet in the organizational mud, it was stretching my capabilities. At the end of the 1980s, I had to choose, and entrepreneurship won for the next eighteen years. Could I start and do something innovative with personal computers myself?
The answer was yes.

Reaching retirement in the 2010s and reflecting on my past experiences and the changes in our world since those 1970s, I wondered what made all this happen. Technological innovation was the phenomenon that fascinated me all along my journey of life. What is the thing we call "innovation"? In many phases of the journey of my life, I tried to formulate an answer: starting with my first book 'Micro-computers, Innovation in Electronics' (1977, technology level), next with my second book 'The Management of Innovation' (1983, business level), and my third book 'Innovation, from Distress to Guts' (1988, society level). In the 2010s I had time on my hands. So I decided to pick up where I left off and start studying the subject of innovation again. As a guest of my alma mater working on my dissertation, I tried to find an answer to the question "What is innovation?"

It started in Delft. And, seen from an intellectual point of view, it will end in Delft.

B. J. G. van der Kooij

B.J.G. van der Kooij

The Invention of the Steam Engine

B.J.G. van der Kooij

Context for the Discoveries

For a person in the pre-steam era, the arrival of the steam engine was a miracle. People of those days were used to manual labor, at home and at work. Work was physical; energy was supplied by humans, water/wind, or animals. Transportation was by foot, on horseback, or in a stagecoach. Travelling was rare, and the world was small for the peasants of those days. The road infrastructure was limited, and the road quality was often bad, especially in winter. Working conditions were not too good, either—especially not in the emerging mining industry, where gasses and water were a continuous threat to life. All that resulted in the preceding ages of mechanization was either human-, animal-, wind-, or water-powered.

Now look at the steam-era person faced with a lot of changes as a new kind of machine emerged. Smoking, hissing, smelling, and burning, a new revolutionary device brought power to the people. A devilish machine started to appear on the streets, called "the puffing devil." Work in the factory changed, too, as tools and machines became powered from a central steam engine—not any more powered by human energy, but through a system of line shafts and belts. Suddenly energy-intensive industries were not any more depending on streams for waterpower. Now factories went to where the people were. It was the start of urbanization and industrialization.

But that was not the only thing that changed. The person in the steam era was travelling differently. Travel on horseback or coach, and the transportation of goods transported by draymen on their horse pulled flat-bed wagons, it all changed considerably when steam power emerged. Travel now went by steam-powered coach, a steamboat, or even by steam-powered locomotives on a rail infrastructure. A businessperson could travel in relative comfort, although the smoke-belching steam locomotives could be a nuisance. That was a small thing compared to the shit-smelling streets in the days of horse-powered transportation or the unreliable ferries. Now the steam-powered ferry could offer its service independently from wind conditions. And as a welcome side effect, coal prices dropped in the city as coal was transported steam-powered over rail and by canal. Heating and cooking did not depend any more on wood and charcoal. That was progress for the people of those days.

So it is not too bold to observe that society changed between those two moments in time due to technical changes initiated by the new phenomenon of steam. It took quite some time, many scientific discoveries, and a lot of engineering effort before this all came to happen within the context that existed in the nineteenth century.

This case study[3] describes the developments that resulted in the steam engine. It covers a range of developments that have to be considered in the context of its time and place. The time frame for these developments was the seventeenth and eighteenth centuries. The place was England. The context was European. It is a story about the madness of times and the creativity of individuals.

The seventeenth and eighteenth centuries

One has to realize the influence of the context of time on the works and successes of the scientists of the seventeenth and eighteenth centuries in Europe. Local and relatively small events—like the religious conflict between Protestants and Catholics—had quite large and supranational consequences. In France the Huguenots were expelled as a result of the Edict of Fontainebleau, issued by Louis XIV in 1685, which declared Protestantism illegal. The Huguenots fled to the Netherlands, England, and Germany. The single event of the edict gave those countries a boost on their path of development.

Turmoil in Europe

But it was not only about religion that conflicts arose, often resulting in wars. It was also about economic, political, and mercantile dominance. It was about expansionism: in turn countries aimed at expanding their territories within Europe or outside Europe (colonialism). An example is the *War of Devolution* (1667–1668), when France expanded into the

---

[3] The content of this case study is not the result of my own *primary* research, but is based on other scholarly work. I have used a broad range of sources, including Wikipedia and sources found through Google Scholar. Where realistically possible these sources are acknowledged.

Spanish-controlled Netherlands and the France-Comte. The same expansionism resulted in the *Franco-Dutch War* (1672–1678), in which France allied with Sweden, the Prince-Bishopric of Münster, and the Archbishopric of Cologne and England, and invaded the Dutch Republic. The *War of the Reunions* (1683–1684), a short conflict between France and Spain and its allies, was based on the territorial and dynastic aims of Louis IV (also called the Sun King). In the east of Europe, the *War of the Holy League* (part of the Great Turkish War during 1683-1699 in which the Ottoman Empire attacked the Habsburg Empire) was a continuation of the religious expansionistic conflicts between Islam versus Christianity. In short, conflicts galore in those days.

Many of the conflicts had a more local cause for the struggle for power: the fight for democracy or religious toleration. In England the *Glorious Revolution* of 1688 saw the overthrow of King James II of England by the Dutch stadtholder William III of Orange Nassau. This bloodless revolution was part of the *Nine Years War* (1688–1697) between King Louis XIV of France and a coalition of Anglo-Dutch Stadtholder-King William III, Holy Roman Emperor Leopold I, King Charles II of Spain, Victor Amadeus II of Savoy, and the major and minor princes of the Holy Roman Empire. The *Great Northern War* (1700–1721) was a conflict between the Tsardom of Russia and the Swedish Empire. The *War of the Spanish Succession* (1701–1714) was fought primarily by forces supporting the French candidate—the Spanish loyal to Philip V of France and the Electorate of Bavaria, together known as the Two Crowns—against those supporting the Austrian candidate, the Grand Alliance: the Spanish loyal to Archduke Charles, the Holy Roman Empire, Great Britain, the Dutch Republic, Portugal, and the Duchy of Savoy. The *War of the Quadruple Alliance* (1718–1720) was a result of the ambitions of King Philip V of Spain to retake territories in Italy and to claim the French throne. It saw the defeat of Spain by an alliance of Britain, France, Austria, and the Dutch Republic. The War of the Polish Succession (1733–1738) was a major European war sparked by a Polish civil war over the succession to Augustus II, King of Poland, which other European powers widened in pursuit of their own national interests. The Seven Years' War (1756–1763) was a world war that involved most of the great powers of the time and affected Europe, North America, Central America, the West African coast, India, and the Philippines.

The efforts by many seafaring countries (for example, England and the Netherlands, but also Spain and Portugal) to open up sea trading as a result of the discovery of new land (by people such as Christopher Columbus, Ferdinand Magellan, Thomas Cooke, Henry the Navigator, and Vasco da Gama) also resulted in conflicts. It was the "Age of discovery," and the European colonial period, starting from the early sixteenth century, resulted in the establishment of colonies in Asia, Africa, and the Americas—colonies that were used to strengthen the home economy.

The Dutch created Indonesian colonies on the "Spice Islands" (Moluccas for cloves, Sumatra for nutmegs, and Timor for sandalwood), trading valuable spices. They even opened up the closed society of Japan by creating a trading post on Dejima, a small fan-shaped artificial island built in the bay of Nagasaki, in 1634. The French colonized the French Caribbean (Haiti) and the Far East (Vietnam); the British colonized South Africa, kicking out the Dutch, and many locations in the Far East (India, Australia) and the Americas (North America, Trinidad, Guiana). The European countries created companies such as the Dutch East India Company (Dutch: Vereenigde Oost-Indische Compagnie, VOC), the East India Company (England), and the Hudson Bay Company (England).

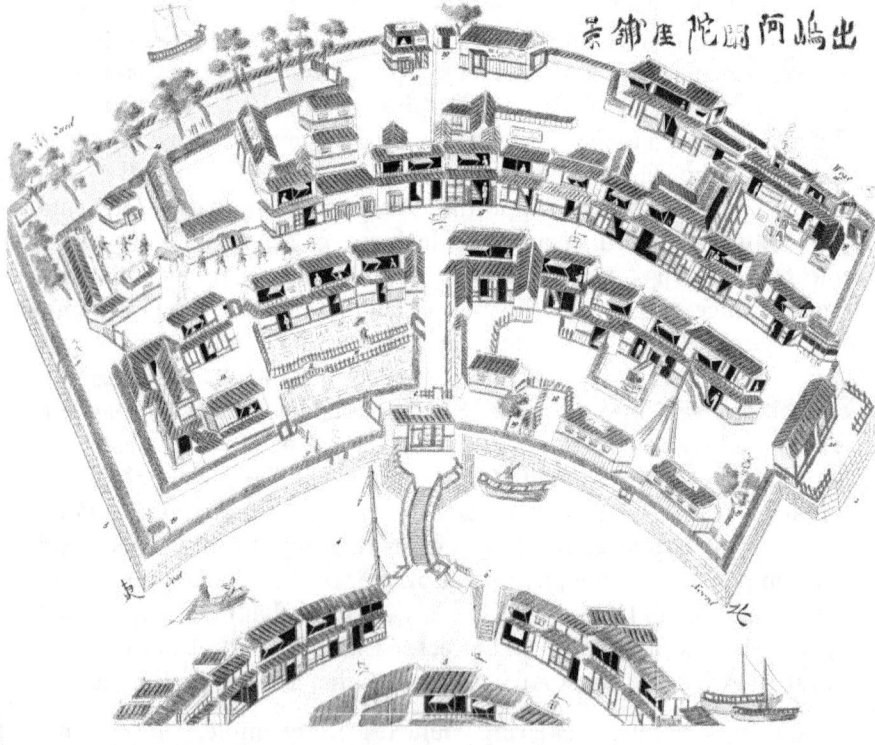

Figure 1: The artificial island of Dejima, Japan (1780).
Source: Wikimedia Commons, Woodblock print by Toshimaya Bunjiemon

These colonial powers fought one another for the best trading posts and routes as the growing competition led to rival nations resorting to military means for control of the spice trade.

Summing up this range of conflicts, one can certainly conclude that the developments that led to the Industrial Revolution took place in turbulent times—a phenomenon to be characterized as the madness of times.

England in the seventeenth and eighteenth centuries

It was not only the madness of times that created a context for the development of societies. There was also the Zeitgeist, the spirit of time; that specific character of a period in time that sets the conditions and limits for man's behavior.

The Spirit of Time

England took part in many of the preceding wars and had its own internal problems based on a range of political, religious, and democratic factors. The following are just a few among those factors that illustrate England in this time frame.

Colonialism and mercantilism: In the years before the eighteenth century, England had become a formidable colonial power, with thousands of its inhabitants colonizing the New Americas and the East. The seas were crowded with English entrepreneurs who extended the range of their business around the globe. The Royal Navy, with its maritime power over the world seas, was absolute; only those annoying Dutch spoiled the fun. For example in 1667, during the second *Anglo-Dutch War*, the Dutch Navy under command of Admiral Michiel de Ruyter on his battleship 'De Zeven Provinciën' sailed to Chatham and "raided the Medway," capturing the fort at Sheerness and threatening the large naval base. In the battle they succeeded in setting fire to three capital ships and ten lesser naval vessels. Then they towed away HMS *Unity* and HMS *Royal Charles*, pride and flagship of the English fleet, as a war trophy. The raid, being a serious blow to the reputation of the English crown, caused a panic in London. But it also helped bring the Second Anglo-Dutch War to an end. The total loss of the Royal Navy of the capital ships must have been

Figure 2: The Medway Raid by the Dutch fleet (1667).
Source: Wikimedia Commons, Jan van Leyden

close to £200,000.[4] Total losses for the Dutch were eight spent fireships and about fifty casualties.[5]

Power, religion, and politics: The political conflicts between the King and Parliament about who was to hold supreme politic power had been fought. The *English Civil Wars*, armed conflicts between Parliamentarians and Royalists, were concluded around 1650 with the Parliamentarian victory. In the *Glorious Revolution,* King James I was overthrown by the Parliamentarians. The United Kingdom of Great Britain, combining England, Scotland, and Wales, was created with the Treaty of Union (Scotland: 1707, Ireland: 1800). Still, this Parliamentary democracy was a continuous clash between the established power of the aristocracy and royalty versus the emerging bourgeoisie and working class. Add to that the religious conflicts with the powers of Rome that had led to the rise of the Church of England separating from Rome during the *English Reformation*. The confrontation of the Anglican

---

[4] To give an impression of the current value of the amounts mentioned, the facilities for recalculation offered at http://www.measuringworth.com were used. This source will be used throughout the case study without further citation. The amount mentioned here would be in 2010 equivalent to £357 million using average earnings.

[5] The author of this case study is Dutch. The metal stern piece (also called transom) of the Royal Charles, showing the English coat of arms with a lion and unicorn and the inscription Dieu et mon droit, is still today on display in the Rijksmuseum in Amsterdam, the Netherlands. On 14 March 2012 the transom was transported to England on board the Royal Netherlands Navy patrol ship *Holland*, accompanied by the Dutch crown prince Willem-Alexander, where it was put on display.

church with nonconformists (i.e., Baptists, Quakers, Congregationalists, and Methodists), however, was dominating English politics. The Test and Corporation Acts (1661, 1673, 1678) had excluded the dissenters from public office and education. This created decades of unrest till the repeal by the Sacramental Test Act of 1828. The dispute over the supremacy over the sea, facilitating colonialism and mercantilism, resulted in wars like the *Anglo-Dutch wars* for control over the seas and trade routes. Not much after the Raid on Medway, the Protestant Dutch prince William van Orange (1650–1702) landed at Brixham in southwest England on November 5th, 1688. He became King of England in 1688 and reigned with his wife, Mary II, till his death in 1702. It was this William III of England who encouraged the passage of the Act of Toleration (1689), which guaranteed religious toleration to certain Protestant nonconformists.

Death, fire and diseases: The frequently returning plague had devastated the country and cities. For example, the *Great Plague of London*, in 1665, killed more than one hundred thousand people (more than 20 percent of the population). This was followed by the Great Fire of London (2–5 September 1666). The material destruction has been computed at 13,500 houses, 87 parish churches, and many other buildings. The monetary loss was estimated to be over £10,000,000 (equivalent to £18,3 billion using average earnings). These catastrophes resulted in social and economic problems that were overwhelming.

Living and working

Within this macro context of power, politics, religion, and natural disasters, people in England lived and worked to survive.

City life: Britain was populated by fewer than 9 million people around 1800. About 3 million were living in the countryside, 1 million in greater London, and the rest in towns and villages. In big cities such as London, the narrow streets were crowded by horse-driven carriages transporting goods and people. Traffic congestion, the loud clatter of horseshoes and iron-rimmed wheels, and the smell of manure were the characteristics of those days (Turvey, 2005, p. 38). Around 1800 London might have been one of the biggest cities of Europe, center of the British Empire, but it was noisy, filthy, and dangerous to live in. The villages and parishes that were within easy walking distance or along the river Thames were supplying the fruits of their efforts to the markets of

London. Those good old times were not that good at all. In the eighteenth century, probably half the population lived at subsistence or bare survival level (Sweet, 1999). In 1700, life expectancy at birth in prosperous England—after the Netherlands the richest country in the world at the time—was only thirty-seven years (Cutler, Deaton, & Lleras-Muney, 2006, p. 99).

Country life: England and Scotland at the end of the eighteenth century were agricultural-dominated countries. The countryside was covered with villages, hamlets, cottages, and farms. Most farmers were smallholders renting up to eight hectares of land and were dependent on raising livestock and dairy farming. The work environment in the so-called "family economy" was dominated by cottage industries. Textiles, for example, were spun and woven in the countryside at home on a large scale. Traders brought the wool and cotton, and the spinners and weavers made the cloth, which was in turn traded by the merchant. Life in the countryside was not easy; survival with the limited food supplies for humans and animals, especially in the winter or after a bad harvest or devastating war, could be problematic. Feeding humans and animals from the depleted soils was a problem. Luckily in the eighteenth century, an agricultural revolution took place in England. Better methods of planting and harvesting and new forms of crop rotation resulted in higher yields, feeding people year-round.

Early mechanization of the manufacturing of goods was starting: the spinning of cotton with the spinning wheel changed when the spinning jenny was developed; weaving was done much more efficiently on the weaving machine. Machines were powered by water (water-driven mills), horses, or men, women, and children. So, workshops where textiles were worked on had to be located near water sources to drive the machines by waterwheels that rotated on the force of water. In addition to that, there was the problem of transportation, for example, for raw materials such as (iron) ore and coal.

Figure 3: Early mechanization (Hargreaves improved spinning jenny).
Source: Wikimedia Commons

Land and water transportation

England was rich in minerals (tin, copper) and coal. The mining of the ore, the transformation to iron, and the final use of it all had a big common problem: the transportation of the raw materials from their source to their destination of use. Traditionally the transport infrastructure available consisted of roads, rivers, and the sea for coastal shipping.

> *Transport by water was highly attractive in an era in which the movement of coal, lime, and other heavy materials by land could require the use of strings of 30 or more pack-horses on roads that were often scarcely passable; between Preston and Wigan, in 1768, Arthur Young found "ruts four feet deep, floating with mud." Sea-going or river vessels were used where possible, with the canals evolving out of the processes of river improvement* (Arnold & McCartney, 2011, p. 217).

So there certainly was a need for a transportation infrastructure that was able to transport large volumes of materials. Next to the road infrastructure, the infrastructure of waterways and canals offered an opportunity. It was time for the canal age and the transportation revolution.

> *The canal age itself dates from 1755; work on Sankey Brook, a tributary of the Mersey, meant that coal could be carried by water from St Helens to Liverpool and led to the promotion of the Bridgewater Canal from the Duke of Bridgewater's collieries at Worsley into Manchester. The success of the small, early canal schemes encouraged more ambitious promotions, as operating on a larger scale could bring "huge savings in manpower and horsepower" and markedly lower (by as much as two-thirds in some cases) the cost of transporting heavy freight, particularly coal.* (Ibid.).

There was, however, one important organizational difference between the waterways and the canals: the former had been generally subject to the jurisdiction of public authorities, city corporations, commissioners, or conservancy boards, whereas canals had been generally constructed and were owned by companies that worked them with a view not only to maintenance, but to profit.

> *Canal companies were normally created with limited liability and sought capital in stages during the period of construction, which could last five to 10 years. They could tap the savings of rentiers,*

as long as they were managed primarily as profit-making concerns; although investors might be "friends, family or a well-disposed banker or merchant," increasingly the canals became the "product of corporate enterprise supported by local shareholders."...During the "mania" period of 1789–96, canal shares were quite widely traded on the London Stock Exchange, and soon became standard investments. In 1811, canal shares represented the largest group of equity shares on the official list of the Exchange, with nearly half the paid-up capital of the equity sector. Across the period 1760–1830, canal construction increased the length of the inland navigation system in England and Wales from 1482 to 3969 miles. Contemporary observers of the new system of inland navigation enthused that "nothing seems too bold" for it to take on and, without a "durable check to national prosperity, its future progress is beyond the reach of calculation" (Arnold & McCartney, 2011, pp. 217-218).

Poor plebs and rich gentry

As poor as the peasants in the countryside may have been, so rich were the members of the ruling class: from baron to duke. The British aristocracy had accumulated, either by inheritance, marriage, or otherwise, large tracts of land that made them rich and let them rule the countryside. They owned the lands, even villages, and got their share of the revenues from exploitation (such as the king, who got—later—his 10 percent share of the patent royalties). But that was not the only way they accumulated their wealth.

Take the 1st Duke of Chandos, James Brydge (1674–1744). He was a Member of Parliament from 1698 to 1714 and, in 1707, had been appointed Paymaster-General of the Forces Abroad, a lucrative office which he held until 1712. During this period £24 million of public money passed his hands. It was common practice to extract "commissions" and "presents" from regiments and contractors which was calculated to yield him more than £716,000.[6] (Dickson & Beckett, 2001, p. 313).

In 1711, the House of Commons launched an enquiry which found a lot of money missing, which Brydges blamed on accounting difficulties. No action was taken against him. He managed to become a Duke: the first duke of Chandos. With all this money he bought land and real estate. He acquired the Cannons, rebuilt it

---

[6] Equivalent to £3.3 billion in 2010 calculating average earnings.

and spent £160,000[7] on it. In March 1721 he bought the manor of Bridgewater. Next to a range of hasty acquisitions in real estate, he became active on the French and English stock markets investing in South Sea Company, Mississippi, African Company and East India Company stock. In total more than a million pounds, funding it by large loans and mortgages. When the South Sea Bubble busted he lost more than £200,000.[8] (Dickson & Beckett, 2001, p. 319).

As his other business activities (like mines in Staffordshire) failed also, he was forced to take mortgages on his real estate and sell part of it. When he died in 1774 he left a financial mess to his heir (Dickson & Beckett, 2001, pp. 333-334).

Mining and water

In the midst of the eighteenth century, a transition had taken place, and a largely agrarian society was transformed. It was a transformation where physical power—wind power, waterpower, human power, and horsepower—was replaced more and more by the "power of fire," a transition that included the change in the primary source of energy: from wood and other biofuels to coal. And coal was to be found in the southwest of England, the areas of Cornwall, around Gwennap and St. Day and on the coast around Porthtowan, and Devon which were among the richest mining areas in the world. The coal mines of Northumberland and Durham, North and South Wales, Yorkshire, Scotland, Lancashire, and other areas supplied the coal for heating and cooking to all those emerging cities in England and Scotland.

Figure 4: British coalfields in 1900s.

Source: Wikimedia Commons

---

[7] Equivalent to £297 million in 2010 calculating average earnings.
[8] Equivalent to £371 million in 2010 calculating average earnings.

Originally mining was restricted to shallow "open pit" mining and working the deposits that reached the surface. That changed when these deposits were depleted and deep mining was needed to access the coals and copper and tin ores. Mining became an important industry. By 1800 Cornwell employed around 16,000 people in seventy-five mines.

> *By 1740 deep mining of copper was underway. The effect of copper mining on Cornwall was huge. Demand for the metal was high, prices were good and copper reserves were large. There was little competition from elsewhere in the country. At its peak the copper mining industry employed up to 30 percent of the county's male workforce and came to involve not just the mining and refining of ore, but also smelting. The county's economic infrastructure was transformed by this industry. Large quantities of ore were moved, mining areas having their entire appearance transformed by the sinking of shafts, the construction of engine houses and the disposal of millions of tons of waste material in surface pits. Ports like Hayle and Portreath were developed and roads, tramways, then railways and even short lengths of canal were built to help move the coal (for the steam engines) to the mines and take away the copper ore for processing.*[9]

Figure 5: Historic mining in Cornwall (Penwith, Geevor, tin mine).
Source: Royal Institution of Cornwall, Geevor, The Tin Mine Museum. www.geevor.com

---

[9] Text from website of Cornwall Heritage Trust (accessed June 2014). Source: http://www.cornwallheritagetrust.org/page_history_industrial_revolution.php.

Working conditions in the coal mines were rather severe. Take the problems of ventilation, particularly as mines became deeper, and (explosive) gases, which were an eternal problem in the (coal) mines. Next were the continuous problems with water from underground streams and waterpockets, making working dangerous when the mines flooded. The conditions were not only severe for the (male) miners, but also for the women and children who were transporting the crushed ore. They were forced to do this because whole families would have to work to get the agreed amount of coal (the "butty system"). In the report from Lord Ashley's Mines Commission of 1842, the following testimony was given by witness No. 26, Patience Kershaw, aged seventeen.

> *My father has been dead about a year; my mother is living and has ten children, five lads and five lasses; the oldest is about thirty, the youngest is four; three lasses go to mill; all the lads are colliers, two getters and three hurriers; one lives at home and does nothing; mother does nought but look after home. All my sisters have been hurriers, but three went to the mill. Alice went because her legs swelled from hurrying in cold water when she was hot. I never went to day-school; I go to Sunday-school, but I cannot read or write; I go to pit at five o'clock in the morning and come out at five in the evening; I get my breakfast of porridge and milk first; I take my dinner with me, a cake, and eat it as I go; I do not stop or rest any time for the purpose; I get nothing else until I get home, and then have potatoes and meat, not every day meat. I hurry in the clothes I have now got on, trousers and ragged jacket; the bald place upon my head is made by thrusting the corves; my legs have never swelled, but sisters' did when they went to mill; I hurry the corves a mile and more underground and back; they weigh 300 cwt.; I hurry 11 a-day; I wear a belt and chain at the workings, to*

Figure 6: A girl pulls a tub of coal (1842).
Source: http://www.bbc.co.uk/, artist unknown

get the corves out; the getters that I work for are naked except their caps; they pull off all their clothes; I see them at work when I go up; sometimes they beat me, if I am not quick enough, with their hands; they strike me upon my back; the boys take liberties with me sometimes they pull me about; I am the only girl in the pit; there are about 20 boys and 15 men; all the men are naked; I would rather work in mill than in coal-pit. (Bourdenet, 2003)

Figure 7: Water flooding a mine (Heaton Colliery in 1815).
Source: Science Museum/SSPL.
http://www.sciencemuseum.org.uk/online/energyhall/page15.asp

The mining of the tin, lead, copper, and coals became more problematic over time as the layers to be explored were located deeper and deeper underground. Getting rid of the water that filled the shafts was a major problem, a problem that caused quite a lot of accidents, with heavy casualties.

A dreadful catastrophe occurred at Heaton main colliery, near Newcastle, by the breaking in of a quantity of water from the old workings, to which the pitman had unhappily approximated too closely...At four o'clock on the fatal morning, Mr. Miller, the resident or under-viewer, visited the men engaged in this operation, and a dripping of water from the roof being pointed out to him, he gave directions that the work should be squared up; and said he would send in the borers to ascertain whether the water proceeded from the waste of the old collieries or not. In less than a quarter of an hour after, the water began to run more freely through the chink; and the two drifters, becoming rather alarmed, sent their boy to apprize two other men who were working near them, with the state of the mine, and to acquaint all the men in the pit with their danger.

The youth, probably impelled by fear, made the best of his way to the shaft, and escaped. The two workmen first mentioned, had now quitted the face of the drift, and presently after, a frightful crash, accompanied by a violent gust of wind, which extinguished the candles, warned them that an immense torrent of water was rushing into the mine; they fled precipitately towards the working-

shaft, distant about a mile; and as the water of course flowed first down the lowest level, reached it just in time to save their lives. The two men who were working near them, the boy just mentioned, and fifteen other men and boys who were on the rolly-way, were so fortunate as to make their escape, but not till the last was up to his waist in water. Every possibility of retreat to those left behind was now cut off; and seventy five human beings, (forty-one men and thirty-four boys) including Mr. Miller, were shut up in the workings towards the rise of the colliery, either to perish by hunger, or to die for want of respirable air. The sufferers who thus found a living grave, left twenty-four widows and seventy-seven orphans, besides Mrs. Miller, and her eight children, to deplore their untimely fate.[10]

This description illustrates the problems with water, but also the problems with foul air and gases. Even more, the transportation and lifting of large quantities of coals and ore within and out of the mines was problematic. In Figure 8 a vertical section of the Dolcoath Mine in Dolcoath, Cornwall, around 1778 is shown, indicating the vertical shafts and the horizontal levels.

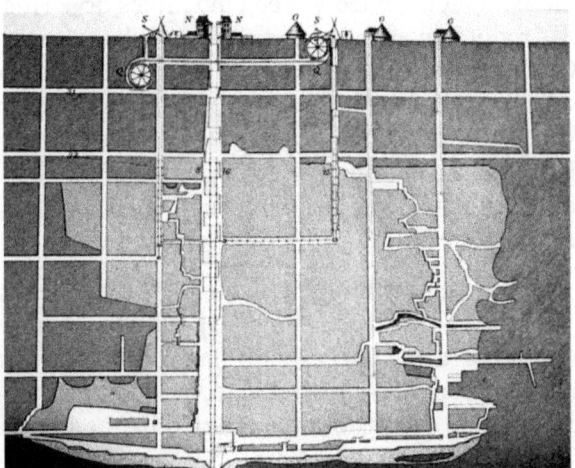

Figure 8: Vertical section of the Dolcoath Mine (c 1778).
Source: (Trevithick, 1872, p. 36)

In the bottom the central drainage area is shown, where the tubing starts for evacuating water. On top is shown the housing for the steam engine (identification "N"). One has to realize that mines like this were important and, economically seen, quite profitable.[11]

---

[10] Source: M. A. Richardson: *Local historian's table book of remarkable occurrences connected with the counties of Newcastle-Upon-Tyne, Northumberland and Durhamy*. Published in five volumes in 1844. http://www.dmm.org.uk/names/n1815-03.htm.
[11] Cook's Kitchen Mine was a very old mine, probably dating back to the seventeenth century and described in 1796 as "one of the most remarkable mines for copper perhaps in the world," although from the 1850s, it used four steam engines and four waterwheels

The mine was a highly profitable concern in its early years and is known to have sold copper ore to the value in excess of £130,000 between 1763 and 1777 (equivalent to £180 million in 2010 using average earnings)...and records show that between 1792–98 the mine sold ore worth £172.246 (equivalent to £182 million in 2010 using average earnings) making a profit of just under £57,750 (equivalent to £60 million in 2010 using average earnings).[12]

A context for change

So, pumping water from the depth of the mines was important. The first mechanization created pumps that were driven by horsepower. Horses already supplied the rotary power needed for hoisting the ore from the depths of the mines to the surface. As the mines became deeper, these methods failed to keep the mines dry. This was resulting in the loss of production and—by the accidents that occurred—the loss of lives. Coal became more and more important as a source of energy, as wood—which was used to create charcoal—became scarce. Coal, for example, was important to fuel the extraction of iron from ore. It also became important to the salt and glass production processes that used a lot of heat. And coal was also replacing wood for household heating and cooking purposes.

It is in this setting that in the eighteenth century in England a period of economic and social change gradually started. After the agricultural revolution came the early mechanization of the textile industries, the first development of iron-making techniques, and the increased use of refined coal. Trade expansion was enabled by the introduction of canals, improved roads, and—later—railways. In the "world of science" of those days, the power of fire was certainly a topic of interest for the gentlemen of science and the engineers of that time. This is the context for the developments that would result in the steam engine.

---

to produce mainly tin. It was also one of the deepest mines. The name is said to derive from a miner named Cook who described the lode he discovered as being as wide as his kitchen. Dolcoath Mine was Cornwall's greatest and longest-lived mine, at the forefront of technical developments and of copper production for much of the eighteenth century, with a workforce of over 1,300 in the nineteenth century. It housed one of the earliest Newcomen engines by 1758 and, working at 917 meters (3,030 feet), was the deepest metal mine in Britain. It finally closed in 1921.

[12] Source: http://www.cornwallinfocus.co.uk/history/cookskit.php; text is referring to "Hatchetts Diary" (Hatchett, 1967).

Science discovers and applies steam

Science was faced with two major hurdles when it tried to understand the mechanism behind the power of fire that in the end resulted in the steam engine. The first was the nature of heat, and the second was the recognition that heat and motion were different manifestations of a wider concept called "energy." The development of steam technology created the need to know more, for example, about the "motive power" of fire. As Sadi Carnot stated in 1824 in his publication *Reflections on the motive power of fire and on machines fitted to develop that power*:

> *Machines which do not receive their motion from heat, those who have for a motor the force of man or of animals, a waterfall, an air current, etc. can be studied even to their smallest details by the mechanical theory. All cases are foreseen, all imaginable movements are referred to these general principles, firmly established, and applicable under all circumstances. A similar theory is evidently needed for heat-engines* (Carnot, 1824, p. 6).

Steam as a phenomenon: power of fire

The power of fire was in different forms interesting for the "gentlemen of science," Not only for its well-known heating properties (industrial and household), but also for other applications, such as the use of gunpowder in military applications (cannons were called "firepots"). So, many scientists of those days were, one way or the other, interested in exploring the power of fire, scientists such as the Frenchman Blaise Pascal (1623–1662), the Dutchman Christian Huygens (1629–1695), the Italian Evangelista Torricelli (1608–1647), and the German Gottfried von Leibniz (1646–1716). After they had more or less discovered the basics of atmospheric pressure and the use of it, efforts were underway to create water pumps using this atmospheric pressure. Otto von Guericke's (1602–1686) work on his air

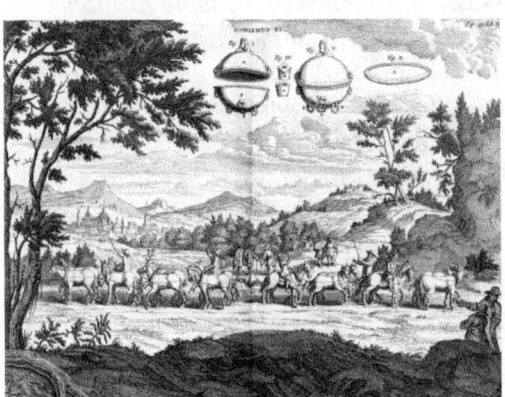

Figure 9: The experiment with the Magdeburg Hemispheres by Otto von Guericke (1654).

Source: Wikimedia Commons, Science Museum

pump—with his well-known experiments with the Magdeburg Hemispheres in 1654[13]—stimulated Robert Boyle (1627–1691) to formulate his law about the volume and pressure of gases (Merton, 1938, pp. 506-512). Huygens worked on a pumping machine using gunpowder and presented a paper on his invention in 1680, "A New Motive Power by Means of Gunpowder and Air."

All these efforts to transform the power of fire into the power of motion were more or less of the engineering type and highly experimental. They gave a certain insight into the application of the power of fire but did not explain it. That understanding started in the eighteenth century, when the gentlemen of science focused their attention on the basic mechanism of our natural environment (for example on the subject of the power of lightning, as described elsewhere,[14]), including the phenomenon of fire and the resulting heat. To understand and to determine what exactly happens when something burns—the nature of heat—was the most pressing issue in chemistry and physics.

Nature of heat: phlogiston theory

The phlogiston theory, a popular theory in those days that had evolved from the late seventeenth century, was still based on the centuries-old concept of the four elements: earth, air, fire, and water. One version, of the many that were developed, was by the German Georg Stahl (1659–1734), professor at Halle, Germany.

> *Developed by the German scientist Georg Ernst Stahl early in the 18th century, phlogiston was a dominant chemical concept of the time because it seemed to explain so much in a simple fashion. Stahl believed that every combustible substance contained a universal component of fire, which he named phlogiston, from the Greek word for inflammable. Because a combustible substance such as charcoal lost weight when it burned, Stahl reasoned that this change was due to the loss of its phlogiston component to the air* (Bohning, 1999, p. 1).

Heat was one of the "imponderable fluids" (like, later, electricity). It was the result of the idea that heat is a fluid that flows from hotter bodies to colder bodies. Joseph Black (1728–1799), a Scottish physician and

---

[13] The Magdeburg Hemispheres, two halves of a ball-like device that could be fitted together, were meant to demonstrate the power of atmospheric pressure by creating a vacuum between the two halves. Two teams of fifteen horses each, connected to the two halves, were supposed to separate the two hemispheres, but they failed to do so.

[14] See the separate published case study in this series of books: The Invention of the Electromotive Engine.

chemist, professor at the University of Glasgow, considered heat to be a substance that could be added to materials. It was supposed to be this matter that expanded bodies when they were heated. He experimented in 1759 to 1763 on "latent heat," and his theory was published in 1803 after his death in *Lectures on the Elements of Chemistry*. It explained his ideas on the different topics of heat: the meaning of heat, the meaning of cold, the nature of heat, and the effects of heat. This theory of latent heat was to be the beginning of thermodynamics, and it stimulated other scholars.

> *After Black's work the investigation of gases proceeded rapidly, most notably in the hands of Cavendish, Priestley, and Scheele, who together developed a number of new techniques capable of distinguishing one sample of gas from another. All these men, from Black through Scheele, believed in the phlogiston theory and often employed it in their design and interpretation of experiments* (Kuhn, 1970, p. 70).

James Watt, the instrument maker who developed and improved the steam engine, was one of Black's students. Discussing with Black his experiments with steam, the link between theory and practical implementation became clear for him.

> *Being struck with this remarkable fact [that steam could heat six times its own weight of water to 212° F], and not understanding the reason of it, I mentioned it to my friend Dr Black, who then explained to me his doctrine of latent heat, which he had taught for some time before this period, (summer 1764,) but having myself been occupied with the pursuits of business, if I had heard of it, I had not attended to it, when I thus stumbled upon one of the material facts by which that beautiful theory is supported* (Fleming, 1952, p. 4).

Nature of heat: caloric theory

Antoine Lavoisier (1743–1794[15]) created a theory by proposing a "subtle fluid" called "caloric" as the substance of heat. Bodies were capable of holding a certain amount of this fluid, hence the term "heat capacity." In 1777 he published *Réflexions sur le phlogistique pour*

---

[15] Lavoisier was a wealthy man, member of the Academy of Sciences. He was executed in 1794 by guillotine in Terror Days of the French Revolution as one of the investors to the tax collectors. These were the hated "fermiers" of the Ferme Générale, an organization of private tax collectors that collected duties on behalf of the king under six-year contracts. The Ferme Générale was one of the most hated components of the Ancien Régime because of the profits it took at the expense of the state, the secrecy of the terms of its contracts, and the violence of its armed agents.

servir de suite à la théorie de la combustion et de la calcination, the first of what proved to be a series of attacks on phlogiston theory. His experiments (Figure 10) changed the way the nature of heat had been explained up to that time: the phlogiston theory was overthrown by the antiphlogistic theory (Conant, 1948).

> *By 1777, Lavoisier was ready to propose a new theory of combustion that excluded phlogiston. Combustion, he said, was the reaction of a metal or an organic substance with that part of common air he termed "eminently respirable." Two years later, he announced to the Royal Academy of Sciences in Paris that he found that most acids contained this breathable air. Lavoisier called it oxygène, from the two Greek words for acid generator... Lavoisier began his full-scale attack on phlogiston in 1783, claiming that "Stahl's phlogiston is imaginary." Calling phlogiston "a veritable Proteus that changes its form every instant," Lavoisier asserted that it was time "to lead chemistry back to a stricter way of thinking" and "to distinguish what is fact and observation from what is system and hypothesis." As a starting point, he offered his theory of combustion, in which oxygen now played the central role* (Bohning, 1999).

Figure 10: The gasometer used by Lavoisier.

Source: Lavoisier, A. L.: Traité élémentaire de chimie. http://historyofscience.free.fr/Lavoisier-Friends/a_tab8_lavoisier_gazometer.html; Wikimedia Commons

At the same time as Lavoisier's experiments, other scientists such as Carl Scheele and Joseph Priestley in the 1770s discovered oxygen and with it the theory of caloric. But despite the discovery of oxygen, the phlogiston theory continued to be accepted. Lavoisier's new system of chemistry (later also called "antiphlogistic chemistry") was in 1789 published in the *Oeuvres: Traité Élémentaire de Chimie* (Elements of Chemistry)

It was Lavoisier who created a revolution in chemistry that would destroy the phlogiston theory, eliminating the four elements of antiquity and replacing them with the modern concept of elements (substances that could not be broken down and that were the fundamental substances of chemistry).

His work would later be referred to as part of the Chemical Revolution (1772–1789).

> *Even Joseph Priestley, the last important defender of phlogiston, admitted that "there have been few...revolutions in science so great, so sudden, and so general, as the prevalence of what is now usually termed the new system of chemistry, or that of the Antiphlogistians, over the doctrine of Stahl* (Siegfried, 1988, p. 35).

In 1824 the Frenchman Sadi Carnot (1796–1832), later often called the father of thermodynamics, published a book *Reflections on the motive power of fire and on machines fitted to develop that power* (Carnot, 1824). It was clear that Carnot was well aware of the importance of solving the water problem in the important British mining industry and was one of the persons who contributed to solving it.

> *Savery, Newcomen, Smeaton, the famous Watt, Woolf, Trevithick, and some other English engineers, are the veritable creators of the steam-engine. It has acquired at their hands all its successive degrees of improvement. Finally, it is natural that an invention should have its birth and especially be developed, be perfected, in that place where its want is most strongly felt* (Carnot, 1824, p. 5).

Carnot stated that motive power is due to the fall of caloric (heat) from a hot to a cold body. He drew a comparison between the work that could be extracted from a waterwheel and that which could be obtained from a steam engine.

> *We can easily recognize in the operations we just described the re-establishment of equilibrium in the caloric, its passage from a more or less heated body to a cooler one...The production of motive power is then due in steam-engines not to an actual consumption of caloric, but to its transportation from a warm body to a cold body. That is, to its re-establishment of equilibrium...* (Carnot, 1824, p. 7).

He explained the relation between heat and motive power by the fall of caloric between a hot body and a cold body and with his model, the *Carnot Cycle*, explained the relation between "thermal energy" (heat) and "motive energy" (work). Despite the fact that the caloric theory of heat was incorrect, Carnot's work brought together insights that remain relevant, and it was used by his successors, leading to the *concept of entropy* in thermodynamic theories of our time.

Nature of heat: frictional heat theory

The nature of heat had its different explanations in theories that changed over time. It was Benjamin Thompson (1753–1814), later Count Rumford, who in 1798 published a paper called "An Experimental Enquiry Concerning the Source of the Heat which is Excited by Friction" that became the starting point of the revolution in thermodynamics. He opposed the caloric theories that heat was a fluid as he had observed the existence of frictional heat[16] when boring a cannon resulted in heating water to a boiling point.

> *Being engaged lately in superintending the boring of cannon in the workshops of the military arsenal at Munich,[17] I was struck with the very considerable degree of Heat which a brass gun acquires in a short time in being bored, and with the still more intense Heat (much greater than that of boiling water, as I found by experiment) of the metallic chips separated from it by the borer. The more I meditated on these phenomena, the more they appeared to me to be curious and interesting. A thorough investigation of them seemed even to bid fair to give a farther insight into the hidden nature of Heat; and to enable us to form some reasonable conjectures respecting the existence, or non-existence, of an igneous fluid [caloric]—a subject on which the opinions of philosophers have in all ages been much divided...It would be difficult to describe the surprise and astonishment expressed in the countenances of the bystanders, on seeing so large a quantity of cold water heated, and actually made to boil, without any fire...And, in reasoning on this subject, we must not-forget to consider that most remarkable circumstance, that the source of the heat generated by friction, in these experiments, appeared evidently to be inexhaustible. It is hardly necessary to add, that anything which any insulated body, or system of bodies, can continue to furnish without limitation, cannot possibly be a material substance: and it appears to me to be extremely difficult, if not quite impossible, to form any distinct idea of anything,*

---

[16] Note the analogy with "frictional electricity," which was observed by scientists trying to understand electricity.

[17] He was a colonel in the British Army, and he held at that moment temporary command of Munich while it was besieged by the French and the Austrians. As a token of gratitude for this and many other contributions to the welfare of his country, the Elector created Thompson a Count of the Holy Roman Empire. The young officer chose the name Rumford in appreciation of the New Hampshire town where he had once been a schoolmaster and where he had won the hand of the widow of the town's most celebrated citizen, the late Colonel Rolfe.

capable of being excited—and communicated, in the manner the heat was excited and communicated in these experiments, except it be MOTION (Rumford, 1798, pp. 81, 92, 99).

Nature of heat: thermodynamic theory

Lavoisier had inaugurated a new era in chemistry by the establishment of the principle of the indestructibility of matter, and now Rumford's idea was the first step towards the equally important law epitomized by the words "conservation of energy."

Then came James Prescott Joule (1818–1889), who examined heat produced by both electrical and mechanical means and was convinced that the various forms of energy could be converted into one another. Joule argued for the mutual convertibility of heat and mechanical work and for their mechanical equivalence. With his paddle-wheel experiments in 1845, he proved that the friction and agitation of the paddle wheel caused heat to be generated in a body of water (Figure 11). After hearing of Joule's ideas, William Thomson, 1st Baron Kelvin (1824–1907), originally in favor of the caloric theory, changed his opinion. Joule and Thomson discussed their different opinions and began a collaboration: Joule conducting experiments, Thomson analyzing the results and suggesting further experiments. The collaboration lasted from 1852 to 1856.

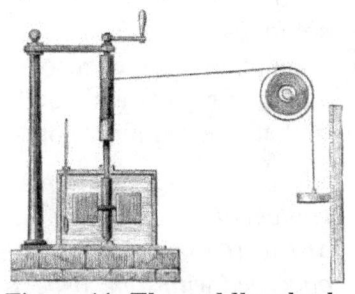

Figure 11: The paddle-wheel experiment: Joule's apparatus for measuring the mechanical equivalent of heat.
Source: Wikimedia Commons

It was the American Willard Gibbs (1839–1903) who in 1875-1878, by publishing parts of the paper "On the Equilibrium of Heterogeneous Substances", formulated the first law of thermodynamics (the conservation of energy), the second law of thermodynamics (the entropy of an isolated system never decreases), and the fundamental thermodynamic relation. He integrated chemical, physical, electrical, and electromagnetic phenomena into a coherent system.[18]

---

[18] For more on this topic, see also (Cheng, 1992).

Steam explored by engineering

It may have been those theories that explained—much later in time—the more fundamental aspects of the nature of heat and its relation to motion, but it was the early efforts of the "hydraulic engineers" that explored the possibilities of applying steam technology. They created steam-based artifacts, sometimes without even understanding the mechanism behind it. On the one hand, these "scientific endeavors" were creating insight; on the other hand, there were efforts to transfer the insights into practical solutions, especially in areas of application that needed solutions for serious problems: military, transportation, and mining—and, not to forget, also in some less serious problems as the entertainment of royalty and nobility. The serious drainage problems of mining in that time certainly got scholarly attention.

> *This all resulted in massive interest in efforts to solve the drainage problem of the mines. Of the 317 patents issued in England from 1561 to 1688, about 75%, (43% directly; 32% indirectly) were concerned with some aspect of the mining industry. It will be noted that 43, or about 14%, of the total of 317 patents were devoted to solving the problem of mine drainage. And about 20% of the inventions patented between 1620 and 1640 were for water-raising and draining devices. This prehistory of the steam engine clearly illustrates the interaction between science and technology, and their preoccupation with specific applications* (Merton, 1938, pp. 502-503).

Denis Papin: vacuum and pressure combined with power

One of those active inventors in those days was the Frenchman Denis Papin (1647–1712), fellow of the important Royal Society, who published in 1688 to 1690 about his ideas for a cylinder with a piston driven by the forces of condensing steam: "Recueil de diverses pièces touchant quelques nouvelles machines." (Papin, 1695). Denis Papin had combined atmospheric pressure and a vacuum to demonstrate motive power. He placed a close-fitting piston into a cylinder, connected to a weight by a cord and pulley. As the piston was raised, the cylinder filled with steam. As the steam cooled, it condensed, producing a vacuum, and atmospheric pressure drove the piston down (Figure 12), lifting the weight that was connected to it. One clearly recognizes the link between heat and motion that he created.

Denis Papin was born in the little village of Chitenay, near the city of Blois in the middle of France. He went to a Jesuit school and then to the

University of Angers, where he became a "docteur." Papin was, together with Gottfried Wilhelm von Leibniz (1646–1716), hired by the Dutch scientist Christiaan Huygens (1629–1695) as a research worker for the Académie Royale des Sciences in Paris. In this capacity he worked on Huygens's idea of using gunpowder to create a vacuum under a piston allowing pressure from the outside air to force the piston down.

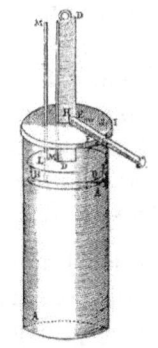

Figure 12: First piston steam engine, developed by Denis Papin (1690).
Source: Wikimedia Commons

> *In 1672, Huygens acquired two young students and collaborators: Gottfried Wilhelm Leibniz, the 26-year-old diplomat, and Denis Papin, a 25-year-old French medical doctor introduced into the Academy by Madame Colbert. Within a year, Huygens and his new colleagues had successfully modified the von Guericke air pump into an engine capable of transforming the force of exploding gunpowder into useful work* (Valenti, 1979, p. 28).

In 1675 Papin, being expelled from Paris, moved to London, where, upon recommendation by Huygens, he obtained a position with Robert Boyle (1627–1691) and may have been responsible for improvements to Boyle's later air pumps. By 1680 Papin made a major breakthrough towards controlling highly compressed steam in the form of his "New Digester for softening Bones, etc.": he developed the steam pressure cooker, and he added a new feature to create a safe device: the safety valve.

> *By this time Papin was in England, collaborating with Boyle in a series of experiments in aerostatics. In 1679, he demonstrated his famous Digester [the early pressure cooker] to the Royal Society. He proposed several plans for raising water from mines. His last suggestion embodied the statement of a practicable method for using atmospheric pressure continuously for the transmission of power over a considerable distance. He likewise suggested the production of a vacuum under a piston by the condensation of steam, stressing in the same memoir the small cost of power thus derived. The uses to which he proposed to put this power reflect the leading economic and technical interests of the day: mining, military technology and shipping* (Merton, 1938, p. 512).

Papin, Leibniz, and Huygens knew one another well. It was Leibniz (1646–1716) who proceeded to discover and develop the science of dynamics and its mathematical tool, differential calculus. He worked on calculating machines and later developed his "vis viva" (living force) theory relating to the measurement and conservation of "force"(Iltis, 1971). So, Papin, Leibniz, and Huygens exchanged ideas and distributed their ideas all over Europe.

Papin left England in 1681 for Venice, where he became a curator for a period of three years. He became director of experiments at Ambrose Sarotti's "Accademia publicca di scienze." After the academy failed for lack of financial support, he returned to England in 1684 and became curator at the Royal Society.

> *In 1687, Papin illustrated the operation of his pneumatic pump by constructing a model fountain. Water was raised by the alternate suction and pressure exerted by a pair of air pumps. Papin enclosed his model in a container, allowing his Royal Society colleagues to observe the water spouting at the top but concealing its internal mechanism, and he then challenged the Royal Fellows to guess at its design. The Royal Fellows failed to solve Papin's puzzle and were especially embarrassed since they all had earlier agreed that the pneumatic transmission of power was impossible. Papin found himself suddenly friendless in London and decided to leave for Germany later that year* (Valenti, 1979, p. 32).

Then in 1688, Papin became professor of mathematics at the University of Marburg.

> *In 1690, Papin published an historic article in the Acta Eruditorum of Leipsig, "A New Method of Obtaining Very Great Moving Powers at Small Cost." Here, for the first time, Papin proposed using the power of expanding steam to operate an engine. In the new invention, steam replaced the gunpowder charge of Huygens's cylinder, creating a more complete vacuum under the piston and thereby taking advantage of the full force of atmospheric pressure* (Valenti, 1979, p. 32).

In 1695 he removed to Cassel, where he assisted his patron, the Landgrave of Hesse, in making experiments upon a great variety of subjects. The Landgrave, who was always involved in the European wars and consequently short of funds, did not lavish resources on Papin. To attract his attention, Papin constantly pursued inventions that would make spectacles. Thus the submarine. The first one failed miserably. The second is said to have made a short trip in the river, but the Landgrave lost interest after the demonstration. Perhaps the main attraction of the steam engine was its potential to pump water into a tank at the top of the palace in order to run the fountains in the garden. Papin named his ultimate steam engine (in 1707) the Machine of the Elector, in honor of Charles-Auguste of Hesse; again it functioned at a demonstration and again the Landgrave lost interest. It is revealing that there was apparently a public demonstration, with the Landgrave at center stage, for every invention.[19]

Leibniz and Papin corresponded through the years, discussing the application of steam to create "work". Leibniz was a great supporter of Papin's efforts. He wrote in 1704:

> *Yet I would well counsel [you], Monsieur, to undertake more considerable things which would force [forcassent] everyone to give their approbation and would truly change the state of things. The two items of binding together the pneumatic machine and gunpowder and applying the force of fire to vehicles would truly be of this nature* (Valenti, 1979, p. 37).

Papin answered:

> *I can assure you that, the more I go forward, the more I find reason to think highly of this invention which, in theory, may augment the powers of Man to infinity; but in practice I believe I can say without exaggeration, that one man by this means will be able to do as much as 100 others can do without it…Yet it's a great shame that the things from which the Public could derive such considerable usefulness aren't impelled by heat. Because the advantages which this invention could furnish for sea-going vessels alone, without counting those of land vehicles, would be incomparably greater than all expected from the transmutation of metals* (Valenti, 1979, p. 38).

Papin continued experimenting and designed another steam engine in 1707 and described it in a pamphlet called "New Method of Raising Water by the Force of Fire." After staying in Hannover for some years, he went to England and presented a copy of his treatise to the Royal Society on 11 February 1708. His proposal to build a boat powered by a steam engine was rejected, partly due to a criticism raised by Savery.

> *Papin, then at Cassel, submitted with his paper, a request for fifteen guineas to carry out experiments, but the Royal Society, like*

---

[19] Data from the biographic page of Denis Papin in the Galileo Project website. Source: http://galileo.rice.edu/Catalog/NewFiles/papin.html (Retrieved June 2014).

our own, did not hand out fifteen guineas at a time. Instead, the matter was referred to Savery in 1708, and in his letter of criticism turning down Papin's design there is a passage in which he damned the cylinder and piston, saying it was impossible to make the latter work because the friction would be too great! (Valenti, 1979, p. 42).

Nothing was heard from Papin after 1712, the year of the erection of Newcomen's machine. Not even a death notice was found. The only thing found was his last letter, dated 23 January 1712, to Sir Hans Sloan: "Certainly, Sir, I am in a sad case since even by doing good I draw ennemi's upon me, yet for all that I fear nothing because I rely on God Almighty" (H.W. Dickinson, 1947, p. 422).

One last aspect. As it was common for a scientist to be supported by a protector, they sought patronage. Papin was dependent on it for a great part of his life. It was, for example, Charles-Auguste, Landgrave of Hesse-Kassel, who appointed Papin Professor of Mathematics at University of Marburg in 1687 to 1695. Leibniz found himself in the same position. He sought patronage at the House of Hannover and served three consecutive rulers of the House of Brunswick as historian, political adviser, and most consequentially, librarian of the ducal library. Huygens was of a rich and influential Dutch family and did not have to worry about his income; he can be considered as a Dutch gentleman of science.

Much more can be said about these men and their contemporaries, but this description illustrates the conditions many scholars were facing: dependency, jealousy, and the "not invented here" syndrome[20] displayed so well by Savery.

Hydraulic engineers: waterworks for the aristocracy

In the relation among fire, heat, and motion being shaped, the first motion artifacts being constructed, and the mathematics more or less getting started, we find the application of these ideas into the world that many of the scholars were familiar with. The nobility of that time always was interested in something entertaining, be it in arts (e.g., music, theatre) or in science (e.g., waterworks).

So, next to the great need of water and air pumps in mining, there was also another application for these pumps in which the royalty, and thus

---

[20] The reasons for not wanting to use the work of others are varied, but can include fear through lack of understanding, an unwillingness to value the work of others, or being part of a wider "turf war" (Wikipedia, Not Invented Here).

the aristocracy, were highly interested: their waterworks. They employed hydraulic engineers who created in their majestic gardens astonishing waterworks. As the hydraulic engineers needed the aristocracy to fund and support their endeavors, they relied on influential patrons to help to assure government approbation. The aristocracy might have considered scientific progress as an interesting pastime for amateurs of science, but the engineers had to earn a living. This was the case in England but also in Germany and France.

Take the example of the Gardens of Versailles, part of the Domaine Royal de Versailles. During King Louis XIV (1638–1715), these used to have more than fifty fountains, such as the Bassin de Latone, Bassin d'Apollon, and Grotte de Thétys. To supply these with water—the amount of water needed per day for these fountains was not much less than the amount of water used per day in the city of Paris—machines were built in the nearby Marly gardens at the Château de Marly.

The "machine" of Marly was a civil engineering marvel located at the bottom of the hill of Louveciennes, on the banks of the Seine about 12 km from Paris (Figure 13). Louis XIV had it constructed to pump water from the river to his chateaux of Versailles and Marly. The construction lasted 7 years and was inaugurated in the presence of the King in June 1684. It was considered a wonder of the world at the time, and may have

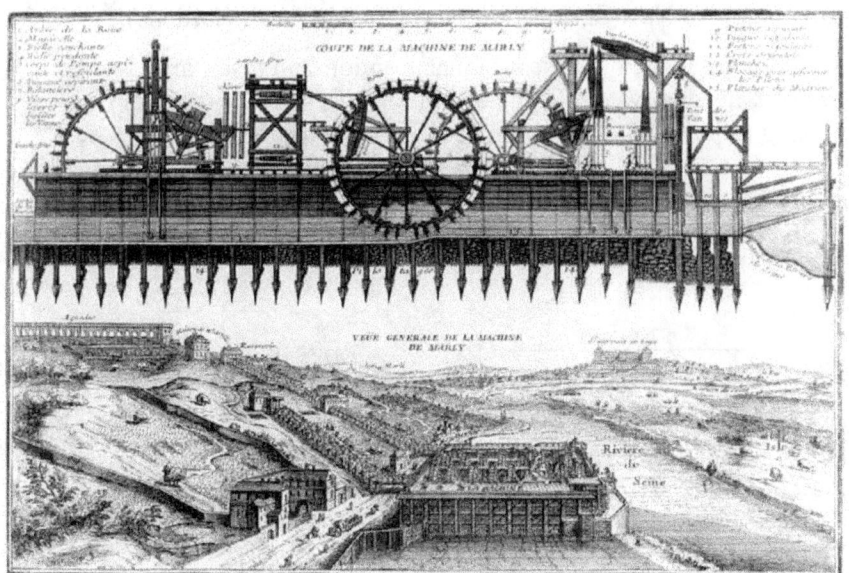

Figure 13: Marly Machine to pump water for the fountains of Versailles, located in the Seine (1684).
Source: www.marlymachine.org/; Wikimedia Commons

been the largest system of integrated machinery ever assembled to that date. Fourteen paddlewheels, each about 38 feet in diameter, were turned by the Seine to power more than 250 pumps, forcing river water up a series of pipes to the Louveciennes aqueduct, a 500 foot vertical rise.[21]

Not only the French needed their entertainment; of course the English nobility was also interested in waterworks and the supply of water to their castles. An example is Edward Somerset, 2nd Marquess of Worcester (1602?–1667), who constructed in his Vauxhall workshop with technician Kaspar Kalthoff, the "Water-Commanding Engine." For this he was granted (by the Royal Ascend) on 3 June 1663 a ninety-nine year patent,[22] with 10 percent of the profits going to the king.

Then there was James Brydges, 1st Duke of Chandos, who lived on the Canons estate with an elaborate water garden. He employed the French-born British priest John Theophilus Desaguliers (1683–1744), who created with his hydraulic engineering efforts fountains for his water garden. And there were the works of Sir Samuel Morland (1625–1695), appointed in 1680 Master to the Works of King Edward II, who was applying his knowledge of mathematics and hydraulics to construct and maintain various machines. Among those were water engines to supply water to Windsor Castle and the gardens in Versailles of French King Louis XIV.[23] He invented the plunger pump in 1675 and made improvements to New Spring Gardens (now Vauxhall Gardens).[24]

Last but not least, big cities such as London, but also Paris, Rome, and Berlin, needed water. They needed lots of water for the ever increasing population. In London it was companies like the *Chelsea Waterworks Company*, the *Lambeth Waterworks Company*, and the *Borough Waterworks Company* that were highly interested in using new methods for distribution of water to the numerous households.

---

[21] Louis XIV and the Creation of Versailles, La Machine de Marly—Excerpt from *l'Encyclopédie, ou Dictionnaire Raisonné des Sciences, des Arts et des Métiers de Diderot et d'Alembert*. Source: http://www.marlymachine.org/. Accessed 30 April 2013.
[22] *An exact and true definition of the stupendous Water commanding Engine, invented by the Right Honourable (and deservedly to be praised and admired) Edward Somerset, Lord Marquis of Worecestere*, &e. &c. (Stat, 15 Car. II. c. 12. A.B. 1663.)
[23] Sir Samuel Morland: *Elevation des Eaux, par toute sorte de Machines, Reduite À la Mesure, Au Poids, À la Balance, par le Moyen d'un Nouveau Piston, & Corps de Pompe, & d'un Nouveau Movement Cyclo-elliptique, en Rejettant L'usage de toute sorte Demaivelles Ordinaires* (1685).
[24] In 1786 James Watt would visit, at the invitation of the French government, the Marley Machine.

Applying the power of fire

Looking at the first Industrial Revolution in retrospect, one might be tempted to see all those changes as fueled and stimulated by a never-ending range of discoveries of new techniques, methods, and processes—all those happenings that are called "inventions," the magic creations by individuals.

Some of the inventions are credited to the patrons that stimulated investigative work by others. One early example of nobility interested in invention was Edward Somerset, 2nd Marquess of Worcester. All his endeavors, efforts, and dedications to realize a hydraulic machine (the water-commanding machine) were described by Charles Partington in his book *The century of inventions of the Marquis of Worcester* ("Historical Account of the Fire Engine for Raising Water") (Worcester & Partington, 1825b). It is obvious that he was not the inventor himself (Dircks & Worcester, 1865, p. XIV) but, being extremely rich, was the man backing up the work of others, such as Kaspar Kalthoff.

But there are other, more fundamental discoveries and inventions by people with hands-on experience, for example, the invention of the condenser-based steam engine by James Watt that gave power to the people, literally and figuratively.

Savery's "Miner's Friend" (1698)

It more or less started when Thomas Savery (1647–1712) developed the stationary "fire engine" or "Miner's Friend" in 1698. It was a steam engine where fire was used to produce steam that would create a vacuum. The engine was of the atmospheric type: it functioned due to the injection of cold water into a space full of steam, causing a vacuum. It was mainly used to pump water from mine shafts. The machine was highly inefficient and had problems caused by the high pressure and temperature that the then known mechanical techniques hardly could manage. In addition to that, the engine had to be less than about 6 meter above the water level to be able to function. This

Figure 14: The Miner's Friend (Thomas Savery, 1698)

Source: Captain Thomas Savery: *The miner's friend: an engine to raise water by fire*. S. Crouch (1702), reprinted McCormick (1827)

required it to be installed, operated, and maintained far down in the mine.

Savery, a former military engineer actively tinkering with mechanics, had already invented and patented an arrangement of paddle wheels, driven by a capstan, for propelling vessels in calm weather (British Patent No. 347).[25] His endeavors to secure its adoption by the British Admiralty and the Navy Board, however, failed (Fox, 2007, p. 25). His ideas for the steam machine were certainly influenced by other developments for water-pumping machines and steam-powered machines in that time, such as the "semiomnipotent" and "water-commanding" engine of Edward Somerset (1602?–1667), the 2nd Marquess of Worcester (Worcester & Partington, 1825a, pp. 99-104). Savery, however, by applying the principle of condensing steam and thus creating a vacuum, is credited for creating the working artifact (Farey, 1827, pp. 89-98).

> *If we make a close comparison between Captain Savery's engine and those of his predecessors, it will be in every respect favourable to his character as an inventor; and, as a practical engineer, all the details of his inventions are made out in a masterly style, all accidents and contingencies are provided for, so as to render it a real working engine; whereas De Caus, the marquis of Worcester, Sir Samuel Morland and Papin, though ingenious philosophers, only produced mere outlines, which required great labour and skill of subsequent inventors to fill up, and make them sufficiently complete to be put in execution* (Farey, 1827, p. 108).

Savery then proceeded to demonstrate his machine to King William III and his court, at Hampton Court, in 1698. He also showed a working model of the steam machine in June 1699 to the Royal Society of London, He had certainly understood how to apply the relevant social networks in that time and produced a small book, entitled *The Miner's Friend: An Engine to Raise Water by Fire* (Savery, 1827). In this he addressed the king and the Royal Society:

> *To the King, Your Majesty having been greatly pleased to permit an experiment before you at Hampton Court, of a small model of my engine described in the following treatise, and at that time to show a seeming satisfaction as of the use and power of it; and having most graciously enabled me, by your royal assent to a*

---

[25] *"Navigation Improved; or, The Art of Rowing Ships of all rates in Calms, with a more Easy, Swift, and Steady Motion, than Oars can,"* etc., etc. By Thomas Savery, Gent., London, 1698.

patent and act of Parliament to pursue and perfect the same...It is upon this consideration I am encouraged, with a profound respect, to throw this performance of mine, with the author, at your Majesty's royal feet, most humbly beseeching your Majesty, that, as it had birth in your Majesty's auspicious reign, you will vouchsafe to perpetuate it to future ages by the sanction of your royal approbation, which is the utmost ambition of, may it please your Majesty, Your Majesty's most humble, most loyal, and most obedient Subject, Thomas Savery.

[to the Royal Society] At the request of some of your members, at the weekly meeting at Gresham College June the 14th 1699, I had the honour of working a small model of the engine before you, and you were pleased to approve of it...Your kindness in countenancing this invention in its first appearance in the world, gives me the hope of the usefulness of it will make it more acceptable to your honourable Society as they are the most proper judges of what advantage it may be to the world...(Savery, 1827).

Savery's patent (№. 356, July 2, 1698) came with a fourteen-years protection, but this was extended in 1699 by an Act of Parliament, the "Fire Engine Act," so that it did not expire until 1733. The title page of the patent reads: "*A grant to Thomas Savery of the sole exercise of a new invention by him invented, for raising of water, and occasioning motion to all sorts of mill works, by the important force of fire, which will be of great use for draining mines, serving towns with water, and for the working of all sorts of mills, when they have not the benefit of water nor constant winds; to hold for 14 years; with usual clauses.*"

The Savery engine was not really a steam engine in the way this is meant today, being in essence a form of suction pump in which steam was condensed in a closed vessel and water sucked up into it by the partial vacuum thus caused (Kanefsky & Robey, 1980, p. 171). Savery's machine had two serious problems: it was quite inefficient, using a lot of coal, and it could not pump above about 6m height.

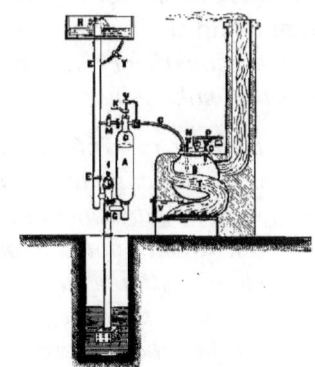

Figure 15: Engine built by Desaguliers (1718).
Source: Robert H. Thurston 1878: *A history of the growth of the steam engine*

Improvement of Savery's pump

Newcomen was not the only one experimenting on Savery's concept. For Desaguliers (1683–1744), born in France, also a member of the Royal Society, the combination of water and fire was also quite challenging. Not only to create the waterworks for James Brydges, 1st Duke of Chandos, but also stimulated by the demonstrations Papin gave at the Royal Society, he created an improvement on Savery's atmospheric pump (i.e., the safety valve).

> *In the engine built in 1718, Desaguliers used a spherical boiler, which he provided with the lever safety-valve already applied by Papin, and adopted a comparatively small receiver—one-fifth the capacity of the boiler—of slender cylindrical form, and attached a pipe leading the water for condensation into the vessel, and effected its distribution by means of the "rose," or a "sprinkling-plate," such as is still frequently used in modern engines having jet-condensers. This substitution of jet for surface-condensation was of very great advantage, securing great promptness in the formation of a vacuum and a rapid filling of the receiver. A "two-way cock" admitted steam to the receiver, or, being turned the other way, admitted the cold condensing water. The dispersion of the water in minute streams or drops was a very important detail, not only as securing great rapidity of condensation, but enabling the designer to employ a comparatively small receiver or condenser* (Thurston, 1878, p. 43).

It was much later in time that William Blakely improved Savery's steam pump and wrote a pamphlet about it: "*A short historical account of the invention, theory, and practice, of fire-machinery: or introduction to the art of making machines, vulgarly called steam-engines.*"

> *Blakely, in 1766, patented an improved Savery engine, in which he endeavored to avoid the serious loss due to condensation of the steam by direct contact with the water, by interposing a cushion of oil, which floated upon the water and prevented the contact of the steam with the surface of the water beneath it. He also used air for the same purpose, sometimes in double receivers, one supported on the other. These plans did not, however, prove satisfactory* (Thurston, 1878, p. 45).

The power of fire understood

After the *experimental scientists* who discovered steam as a force of motive power, and the hydraulic *engineering scientists* who applied this understanding in the practical world, *theoretical scientists* ended up with knowledge about—among other subjects such as electricity, chemistry, and physics—the fundamentals of heat, steam, and motion. Together they managed to create the body of knowledge and managed to use this knowledge and apply the steam technology (Figure 16).

This sounds simple, but one has to realize that many of these latter, and even more fundamental, developments in science took place in a world in turmoil—turmoil in terms of conflicts between church and state. At the beginning of the eighteenth century, England had its religious movements, with nonconformists such as the Baptists, Quakers, Methodists, Unitarians, Puritans, and Presbyterians. The *Test and Corporation Acts* (1673 and 1678), a series of English penal laws that imposed various civil disabilities on Roman Catholics and nonconformists, had also restricted the rights of dissenters. There also

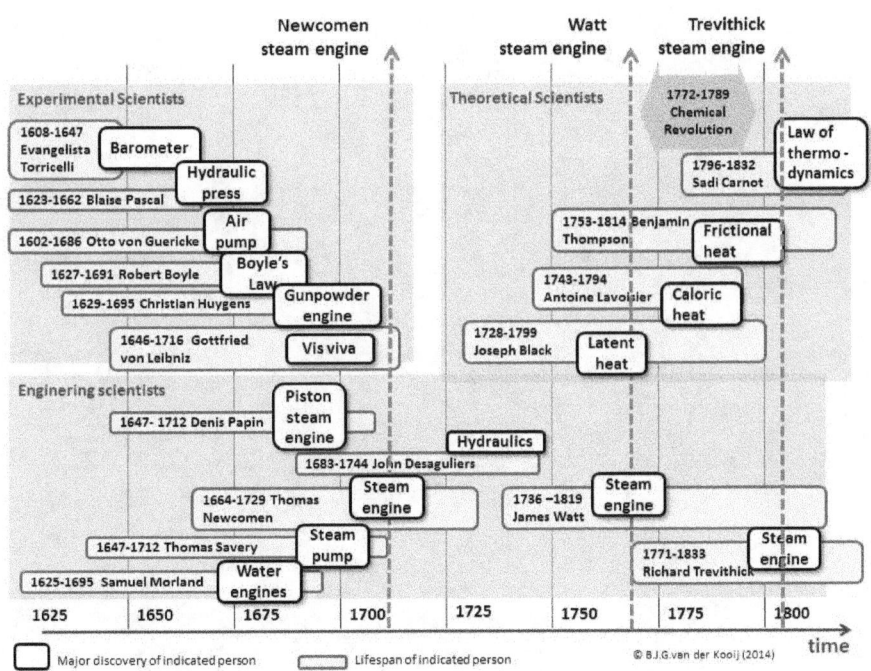

Figure 16: Science discovers the power of steam.
Source: Figure created by author

were social unrests in England related to the effects of the first Industrial Revolution, in which mechanization resulted in massive social changes.

There was turmoil at the end of that century, with political unrest all over. England had its problems with America: from the Boston Tea Party (1773) to the American War of Independence (1775–1783). In France it was the fall of the Bastille in Paris in 1789, leading to the French Revolution until 1794. On top of that were the wars between France and England (1793–1815). All this affected those scientists who lived and worked in that era.

It was at this moment in time, the end of the seventeenth century and the beginning of the eighteenth century, that the foundation of the basic concept of the steam engine was created. The development trajectory would take more than a century, and the three major inventions that were part of that trajectory, would change the world drastically.

First-Generation Steam Engines (1700-1775)

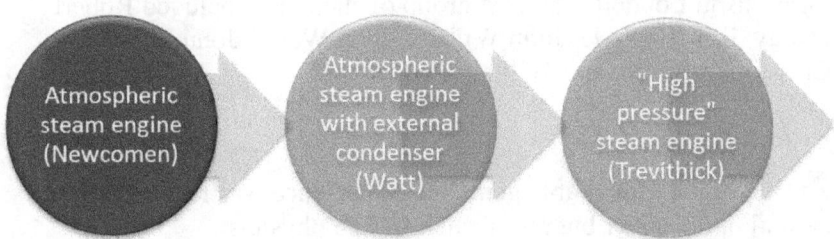

The preceding chapter describes the efforts made by numerous scientists and inventors that slowly led to a cumulating knowledge as a result of scientific curiosity trying to solve real-world problems. It also shows that these efforts have to be seen in the context of the Spirit of Time.

Aristocracy: the gentlemen of science

It was a time when European societies were dominated by aristocracy. It was those nobles, only a small part (less than 2 percent) of the population, who played a dominating role. Being born a noble automatically guaranteed a place at the top of the social order, with all of its attendant special privileges and rights. The legal privileges of the nobility included judgment by their peers, immunity from severe punishment, exemption from many forms of taxation, and rights of jurisdiction. It was also the same aristocracy that furnished from their ranks the scientists who, having the means and time to do so, created the foundations for all those discoveries. Science became fashionable, and it was almost abnormal for "gentlemen of culture" to overlook the "charms of science":

The Royal Society was one of the hobbies of the king. Distinguished personages patronized science often making available considerable sums of money for research purposes, increasing the social reputation of science. Science had definitely been elevated to a place of high regard in the social system of values; and it was this positive estimation of the value of science— an estimation which had been gradually becoming increasingly favorable—which led ever more individuals to scientific pursuits (Merton, 1938, pp. 383-387).

The scientists created their own peer groups, such as the Royal Society, a self-governing fellowship of scientists created in the 1660s, started by physicians and natural philosophers, meeting at a variety of locations in London. The first group of such men included Robert Moray, Robert Boyle, John Wilkins, John Wallis, John Evelyn, Christopher Wren, and William Petty.

Robert Boyle

One of the remarkable gentlemen of science was Robert Boyle, a natural philosopher but also a chemist and physicist.

Robert Boyle (1627–1691) born in Lismore, Ireland, was the son of Richard Boyle, Earl of Cork, an Elizabethan adventurer who enriched himself in Ireland...Robert Boyle attended Eton for four years and then was educated by private tutors, mostly on the continent. He had no university degree. However, he was resident in Oxford for about twelve years, from 1656 to 1668, and he clearly absorbed a great deal of university culture...With his wealth, Robert Boyle became rather a patron himself...He was chronically ill, looked industriously for cures, and wrote and published on this as well.[26]

Boyle was certainly living in privileged circumstances, being the son of the enormously wealthy[27] Richard Boyle, 1st Earl of Cork, the Lord Treasurer of the Kingdom of Ireland. He had private tutors while travelling in 1641 to Europe and staying the winter in Florence.

---

[26] The Galileo Project: Robert Boyle. Source: http://galileo.rice.edu/Catalog/NewFiles/boyle.html.

[27] Richard Boyle, Robert's father, acquired an enormous wealth during his lifetime. From a Kentish yeoman family, Boyle arrived in Ireland in 1588, and by fair means and foul, he made himself the richest man on the island and the first earl of Cork in 1620. It was the time of the colonization of Ireland, just before the Irish Rebellion of 1641. For more on this topic, see: (Canny, 1982), (Ranger, 1957).

> *In 1659 he and Robert Hooke, the clever inventor and subsequent curator of experiments for the Royal Society, completed the construction of their famous air pump and used it to study pneumatics. Their resultant discoveries regarding air pressure and the vacuum appeared in Boyle's first scientific publication, New Experiments Physico-Mechanicall, Touching the Spring of the Air and Its Effects (1660). Boyle and Hooke discovered several physical characteristics of air, including its role in combustion, respiration, and the transmission of sound. One of their findings, published in 1662, later became known as "Boyle's law." This law expresses the inverse relationship that exists between the pressure and volume of a gas, and it was determined by measuring the volume occupied by a constant quantity of air when compressed by differing weights of mercury.*[28]

Over time Boyle, being one of the founders of modern chemistry, would be remembered by Boyle's Law.[29]

European influences

Like Boyle travelled and lived in Europe, other scientists travelled Europe—maybe not in the same style as the rich Boyle, but they lived in different countries, thus spreading their knowledge. These—although often not too wealthy—foreigners, who fled their home country in times of religious unrest and warfaring among nations, could become (foreign) members of the Royal Society. It is interesting to see how the lives of such scientists was dominated by the time frame they were living in. Take, for example, the earlier mentioned Denis Papin[30] in 1680, a man who lived all over Europe, depending on different patronages, and who is considered by some to be the inventor of the steam machine.[31]

> *From July to December 1679 Papin was employed at the Royal Society by Hooke as an amanuensis, and during part of 1680 he*

---

[28] Text form 'Robert Boyle' in Encyclopedia Britannica. Source: Encyclopedia Britannica, http://www.britannica.com/EBchecked/ topic/76496/ Robert-Boyle. (retrieved June 2014)
[29] Boyle's Law describes the inversely proportional relationship between the absolute pressure and volume of a gas if the temperature is kept constant within a closed system.
[30] There were religious reasons why Papin left France. He was a Calvinist, born into a Huguenot family, and after the Edict of Nantes, which had granted religious liberty to the Huguenots, was revoked by Louis XIV in 1685, he became an exile.
[31] He wrote *De Novis Quibusdam Machinis*, a treaty on the steam machine. Full title: *Fasciculus dissertationum de novis quibusdam machinis atque aliis argumentis philosophicis, etc.* (Marburg: J.J. Kürsnerii, 1695). It was translated into French: *Recueil de diverses pièces touchant quelques nouvelles machines, etc.*

was again at Paris with Huygens. He was elected fellow of the Royal Society in 1680, and in 1681 he left England for Venice, where he remained for three years, acting as curator of a scientific society established by Sarotti. He renewed his connection with the Royal Society in 1684, and on 2 April of that year he was appointed curator at a salary of 30l. per annum, his principal duty being to exhibit experiments at the meetings...In 1687 he became professor of mathematics at the university of Marburg, and in 1695 he removed to Cassel, where he assisted his patron, the landgrave of Hesse, in making experiments upon a great variety of subjects. At the end of 1707 he was again in London, endeavoring to interest the Royal Society in his steam-navigation projects, and to induce them to institute comparative experiments of his steam engine and that of Savery...Papin's claims to be regarded as "the inventor of the steam engine" have been advocated with considerable warmth by many French writers, but his labours in this direction have little connection with his career in England, and all the evidence adduced is inconclusive...It is often asserted that he actually made a steam engine, which he fitted in a boat in which he intended to cross the sea to England. It is true that he did construct a boat with paddle-wheels, which was destroyed by the boatmen on the Weser at Münden in 1707; but there is no evidence whatever that the boat was propelled by steam power...From the time of his arrival in England in 1707 he seems to have lived on small payments received from the Royal Society; but all his early friends were dead, and little is heard of him. The date and place of his death are alike unknown.[32]

In this context of the gentlemen of science and the not so wealthy engineers, the next big step in the evolution of steam engines was taken.

Atmospheric engines

The period of the first generation of steam engines is the period in which low-pressure steam engines were developed. Machines were based on the atmosphere vacuum principle, which was proved feasible with Savery's steam pump. However, Savery's steam pump had limited application due to its basic construction. So, solutions to this problem were soon developed. One of the suggestions came from Papin in 1690, as he proposed to use the condensation of steam to create a vacuum

---

[32] Richard Bissell Prosser: *Oxford Dictionary of National Biography, 1885–1900, Volume 43*, Denis Papin.

beneath a cylinder that had previously been raised by steam. However, it was Thomas Newcomen, together with his assistant John Crawley, who created a practical solution for the height problem.

Newcomen's atmospheric steam engine (1712) [33]

Thomas Newcomen (1664–1729) was born in Dartmouth in the southwest of England. He was descended from an aristocratic family that had lost its property during the reign of Henry VIII.

> *Bryan Newcomen was implicated in the 1536 Lincolnshire Rising against Henry VIII and his dispute with Rome. Having incurred the King's displeasure, Bryan's lands were confiscated...Charles Newcomen, younger brother of Bryan, removed to London and it was his two sons, Elias and Robert, who respectively founded the Devonian and Irish family branches. Robert managed to retrieve the family's ancient status when he was knighted at Dublin Castle in 1605 and created Sir Robert Newcomen, Baronet of Kenagh, Co. Longford. This branch of the family died out in the nineteenth century following the death of Viscount Newcomen...Thomas the Inventor was the great-grandson of Elias Newcomen who founded the Devonian branch in 1594* (Corfield, 2013, pp. 211-213).

His grandfather and father were merchants and nonconformist Baptists, and Newcomen followed them in both respects. During the 1680s he became an ironmonger in partnership with John Calley, plumber and glazer and fellow Baptist, who later collaborated with him on the development of the steam engine. Newcomen became a leader of the local Baptists and often preached to their congregations.[34]

> *Thomas was frequently away on business for days, weeks and sometimes months on end...It was during his travels to the Bromsgrove area that brought Thomas into contact with Humphrey Potter who was a member of the Netherton Baptist chapel and where Thomas preached and was invited by Potter to become a trustee of the chapel...Another important associate in later years was Edward Wallin, a Swede and a Baptist minister, who was a member of the group of Proprietors of Savery's Engine*

---

[33] *Oxford Dictionary of National Biography*. Thomas Newcomen, Retrieved on 7 February 2013 from Oxford DNB: http://www.oxforddnb.com/templates/article.jsp?articleid=19997&back=.

[34] Newcomen, Thomas, *Complete Dictionary of Scientific Biography*, 2008. Retrieved 7 February 2013 from Encyclopedia.com: http://www.encyclopedia.com/doc/1G2-2830903150.html.

Patent and who was effectively Thomas's London agent...There is little doubt that Newcomen, in the course of his ironmongery business, would be talking to mine owners among his customers and would be very much aware of the problems of water flooding the mine galleries. Thomas Savery also came to Dartmouth on business and it seems that the two men came to know each other (Corfield, 2013, pp. 214-215).

The combination of his awareness of the problems with the flooding of the mines, his practical knowledge as an ironmonger, and his acquaintance with Savery might have been the breeding ground for Newcomen's invention. He combined the Papin piston, the Savery boiler, and the injecting of water to make a vacuum under the cylinder to create his machine.

The idea of using a jet of water may have been serendipitous. It has been suggested that a hole formed in the cylinder wall through which jacket water penetrated and thereby condensed the steam more rapidly. He connected a beam to the piston rod and the other end to the pump rods, enabling the piston to pull the rods and water up and the weight of the pump rods would then pull the beam down again (Corfield, 2013, p. 217).

In 1712 Thomas Newcomen first unveiled his steam-driven piston engine, which allowed the more efficient pumping of deep mines (Figure 17). Although based on the relatively simple principles developed earlier, Newcomen's engine was a rather complex machine. Early versions of the engine had to be operated manually

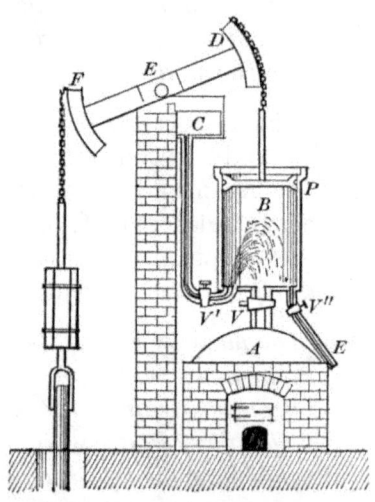

Figure 17: Diagram of Thomas Newcomen's steam engine.

The steam was generated in boiler A. The piston P moved in cylinder B. When the valve V was opened, the steam pushed up the piston. At the top of the stroke, the valve was closed, the valve V was opened, and a jet of cold water from tank C was injected into the cylinder, thus condensing the steam and reducing the pressure under the piston. The atmospheric pressure above then pushed the piston down again.
Source: Wikimedia Commons

opening and closing the valves (no problem, as the speed of the machine was quite slow). In later versions a connection between valves and the rocking beam would open and close the valves automatically when the beam reached certain positions.

> *One thing is very remarkable: as they were at first working, they were surprised to see the engine go several strokes, and very quick together, when, after a search, they found a hole in the piston, which let the cold water in to condense the steam in the inside of the cylinder, whereas, before, they had always done it on the outside. They used before to work with a buoy to the cylinder, inclosed in a pipe, which buoy rose when the steam was strong and opened the injection, and made a stroke; thereby they were only capable of giving 6, 8, or 10 strokes in a minute, till a boy, named Humphrey Potter, in 1713, who attended the engine, added (what he called a scoggan) a catch, that the beam always opened, and then it would go 15 or 16 strokes a minute. But, this being perplexed with catches and strings, Mr. Henry Beighton, in an engine he had built at Newcastle-upon-Tyne in 1718, took them all away but the beam itself, and supplied them in a much better manner* (Thurston, 1878, p. 61).

The force of the engine was limited to mere atmospheric pressure, and the design limited to raising water from mines, and the machine was alternately cooling off and heating up the same cylinder, wasting tremendous amounts of steam and consuming massive quantities of fuel.

> *These chains of inspiration continued through the very end of the inventive process. It is plausible, for instance, that Savery knew and was inspired by the work of Porta, de Caus, and Somerset, since elements of their devices were incorporated in his engine.40 The connections between Newcomen and his predecessors, particularly Papin, are more tentative. It is possible that Newcomen became acquainted with Papin's 1690 device through the latter's publications in Latin and French via the influx to Devon of Huguenot refugees following the Revocation of the Edict of Nantes. Baptist preachers, Newcomen being one of them, were known for their proclivity toward learning. The review of Papin's piston-and-cylinder device in the Philosophical Transactions was another possible channel of information, though one may point out that scientific work emanating from London rarely trickled down to remote Devonshire. And it has been claimed that Newcomen may have become acquainted with Papin's work through contacts with Hooke, though this claim has been discredited.41 There is no*

> *doubt that there was a huge leap between Papin's little experimental device and the complexity of Newcomen's engine, the very first one to generate economic benefits; no matter, however, the level of ingenuity involved in the latter, it is hard to fathom the possibility that Newcomen knew nothing of the conceptualizations of Guericke and Papin* (Kitsikopoulos, 2013, pp. 339-340).

Newcomen's engine was a combination of many preexisting elements and a true atmospheric engine in that it used steam only to create a vacuum and utilized the power of the atmosphere pressing upon the piston to do all the work. Newcomen's model did exhibit conceptual flaws such as the spraying of water inside the cylinder full of steam to achieve a rapid condensation leading to a waste of steam power. But it was a dramatic improvement over Savery's model and proved to be the first engine in history of sizeable economic benefits by providing the definitive solution to the mine flooding problem.

As Savery's patent covered all engines that raised water by fire, Newcomen was forced to go into partnership with Savery because Savery claimed that Newcomen's work was a modification of his own work. And the Act of Parliament of 1699 was quite specific at this point: "No person he or his assigns may make, imitate, use or exercise any vessels or engines for raising water or occasion motion of any sort of millworks by the impellent force of fire" (H.W. Dickinson, 1939, p. 42). By 1712 arrangements (i.e., payment of royalties) had been made with Savery to develop Newcomen's more advanced design of steam engine, which was marketed under Savery's patent. Soon his "atmospheric" engines had been installed in most of the important mining districts of Britain: mines in the Black Country, Warwickshire, and near Newcastle upon Tyne; at the tin and copper mines in Cornwall; and in lead mines in Flintshire and Derbyshire.

After the death of Savery in 1715, his patent was vested in a company called *The Proprietors of the Invention for Raising Water by Fire,* which issued licenses to others for the building and operation of steam engines (A. Nuvolari et al., 2003, p. 5). The licensee fee could be as much as £420 per year patent royalties (equivalent to £737,000 or about €850,000 in 2010 using average earnings calculation) for the construction of steam engines (Oldroyd, 2007, p. 14). This steep pricing policy certainly did not prohibit the widespread use of the engine. Some 104 had been installed when the patent-protection ended in 1733.[35] Newcomen, however, did not derive too much pecuniary advantages of his work. He died in 1729.

Newcomen's steam engines, combined with a pump mechanism, being robust and relatively simple and being able to operate in deeper mines, were a big improvement over the Savery pump. However, they still consumed a lot of fuel. This was one reason why they were not that popular in regions where coal was expensive. Only in those regions that had coal mines were a lot of Newcomen's machines being built. An area that, over time, became larger, as the machines gained in popularity (Figure 18). As the linear (vertical) movement made rotary applications difficult, they were not used for other applications.

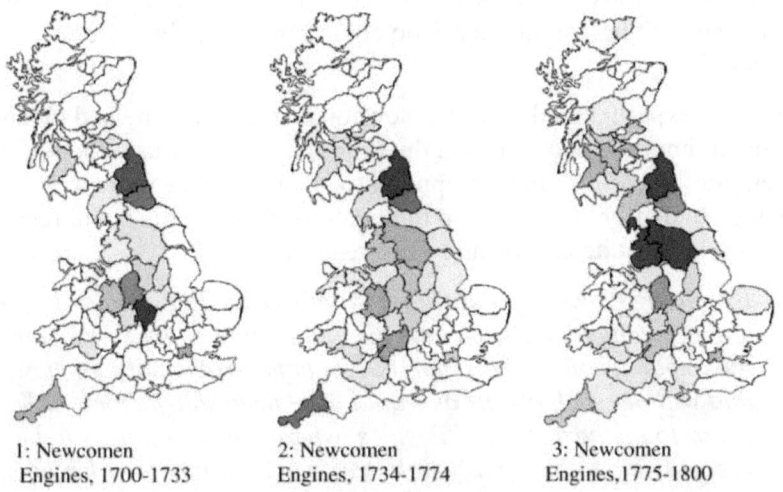

1: Newcomen Engines, 1700-1733
2: Newcomen Engines, 1734-1774
3: Newcomen Engines, 1775-1800

Figure 18: Geographical diffusion of Newcomen steam engines (1700–1800)
Legend: Color from light grey to dark grey indicates an increasing number of steam engines erected. Source: (A. Nuvolari, Verspagen, & von Tunzelmann, 2003, p. 6)

---

[35] *Oxford Dictionary of National Biography*, Thomas Newcomen. Retrieved on 7 February 2013 from Oxford DNB: http://www.oxforddnb.com/templates/article.jsp?articleid=19997&back=.

Improvements of the Newcomen steam engine

Over time the Newcomen machine was improved by engineers such as John Smeaton and John Curr of Sheffield. It concerned modifications on the original concept: for example, another shape for the piston (Smeaton) or a more remote boiler location (Curr).

John Smeaton's steam engine

John Smeaton (1724–1792), an English civil and mechanical engineer, was well known for his work on waterwheels. In 1759 he had published *An Experimental Enquiry Concerning the Natural Powers of Water and Wind to Turn Mills, and Other Machines, Depending on a Circular Motion.*[36] This contribution was the first time waterwheels had been treated seriously in the scientific community (as compared to the strictly engineering community, who could build them but rarely analyzed them).

He also experimented with the Newcomen steam engine and made significant improvements around the time James Watt was building his first engines. In 1778 Smeaton applied to Watt for a license to attach the patented condenser and air pump to the atmospheric engine, and received the following—quite cryptic and evasive—reply:

> *By adding condensers to engines that were not in good order, our engines would have been introduced into that county (which we look upon as our richest mine) in an unfavorable point of view, and without such profits as would have been satisfactory, either to us, or to the adventurers. Besides, where a new engine is to be erected, and to be equally well executed in point of workmanship and materials, an engine of the same power cannot be constructed materially cheaper on the old plan than on ours. The idea of condensing the steam by injecting into the eduction-pipe, was as early as the other kinds of condensers, and was tried at large by me at Kinneal. We shall have four of our engines at work in Cornwall this summer; two of them are cylinders of 63 inches diameter, and are capable of working with a load of 11 or 12 lbs. on the square inch* (Farey, 1827, p. 329).

Was it a positive or a negative reaction? In the years to come, despite Watt's lack of cooperation, Smeaton designed some forty-nine watermills, another six watermills with returning engines (steam pump

---

[36] *Philosophical Transactions of the Royal Society*, vol. 51 (1759–60), pp. 100–74.

used to raise water that in turn drove a waterwheel and thus the machinery), six windmills, and two horse-powered mills.

A cluster of innovations

In the preceding we have identified—as far as possible through the fog of time—the efforts of all those inventive men who struggled and fought for their ideas and who were depending on others to realize them. This dependence could be from the establishment of those days (royalty and nobility). They were the powers that issued protection by patents, decided upon an item's use in military affairs, or funded the projects by ordering those exciting new devices to show off their glorious power. Or it could be from their clients, in many cases, mine owners having a problem with water and air quality in mines.

But whatever the contextual issues may have been, they all created the infrastructure in which the hydraulic engineers and scientific experimenters operated. Building on the principles originating from men such as Papin and Savery, it was Newcomen who created the engine that had impact—impact in terms of the number of machines built and used.

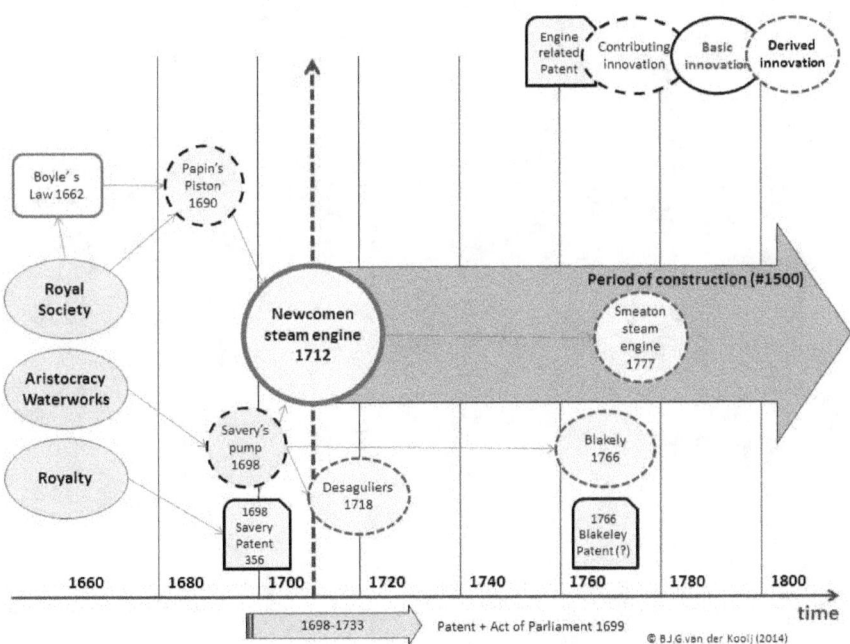

Figure 19: Cluster of innovation around Newcomen's engine.
#1500: Over time, up to the early 1800s, there would be constructed about 1500 engines.
Source: Figure created by author

It was a design that would be operational for quite some time (Figure 19).

As steam engines were large stationary constructions, with a lot of construction for housing and a lot of individual engineering, all engines were in a way individually different. They all, some 1,500 that were built in total, had one characteristic in common: they were quite fuel-inefficient. That was not too much of a problem in regions where coal was in abundance available within short distances. But when that was not the case, the people who wanted steam power had to wait for another type of steam engine. And that would take another fifty years.

Second-Generation steam engines (1775-1800)

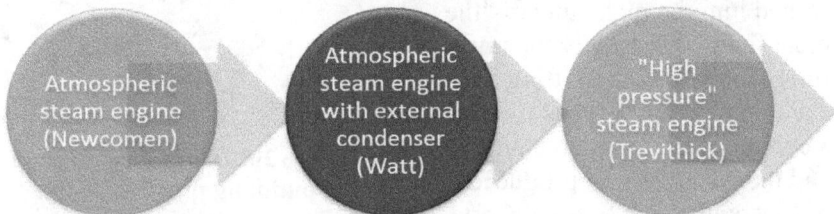

England in the first half of the eighteenth century was immersed in a range of developments that would—later—be called the (first[37]) *Industrial Revolution*. As wood was replaced by coals as a source of energy, mining became important. All the problems related to mining did result in a lot of discoveries, as seen in the previous chapter. Take, for example, coal that was used to melt iron ore more efficiently in iron foundry works. Here a range of discoveries created new ways to produce coke "pig iron"[38] (coke pig iron was cheaper than charcoal pig iron) and later "cast iron."

These developments were important, as it was the interaction where one range of developments influenced another range of developments, resulting in improvements to be used in the first "range." To clarify this, one only has to look at the development of the steam engine. Steam engines were used in the steel mills and foundries, thus making it possible to create cast iron. Cast iron, in its turn, was used to create parts for the steam engines. For example, the more accurate cylinders were

---

[37] Sometimes the period of mechanization is called the "first" Industrial Revolution.
[38] In a blast furnace, a blast of air is passed through a charge of iron ore, limestone, and some form of carbon. This "smelting" process is a chemical change: the molten iron produced is tapped from the furnace as "pig iron."

used for Watt's steam engine.[39] This made it possible to create better steam engines that could be used in the (steel) mills and other stationary applications.

Next to the developments in mining, development in metallurgy became important. Many inventors contributed, and among all those participants was Henry Cort. He was experimenting with the process to convert cast iron to wrought iron, which needed the removal of carbon impurities. The process he developed, called "puddling," earned him the nickname of "the great finer." Originally the favorite fining method was "potting and stamping," using coal as fuel: it was a complex and expensive process, and the wrought iron produced was of low quality for forging. Cort developed the puddling furnace, an adaptation to a type of furnace used for casting iron, the air furnace (Figure 20). He patented the puddling process in 1784.

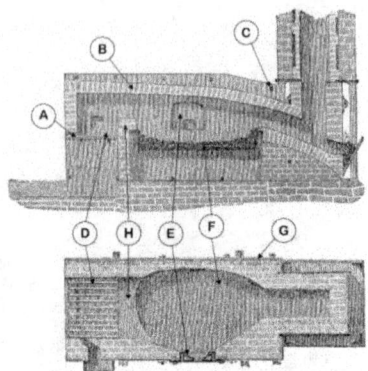

Figure 20: Henry Cort oven for the puddling process (1780).
A. Fireplace grate; B. Firebricks; C. Cross binders; D. Fireplace; E. Work door; F. Hearth and Bottom; G. Cast-iron retaining plates; H. Bridge wall.
Source: Wikimedia Commons

> *Henry Cort, having accumulated capital by serving 10 years as a civilian official of the Royal Navy, bought an ironworks near Portsmouth in 1775. In 1783 he obtained a patent for grooved rollers that were capable of producing iron bars more quickly and economically than the old methods of hammering or of cutting strips from a rolled plate. The following year he patented his puddling process, which consisted of stirring molten pig iron on the bed of a reverberatory furnace (one in which the flames and hot gases swirling above the metal provide the heat, so that the metal does not come in contact with the fuel). The circulating air*

---

[39] It was John Wilkinson, a great exponent of cast iron, who, amongst other things, cast the cylinders for many of James Watt's improved steam engines (till Watt started to make the cylinders himself in the Soho foundries): "In 1775 Wilkinson made his first steam engine cylinder for the firm of Boulton and Watt, which proved very satisfactory when other ironmasters had been unable to meet Watt's specifications." Source: J. R. Harris, "John Wilkinson" (1728–1808), ironmaster and industrialist. www.oxforddnb.com/view/printable/29428.

removed carbon from the iron. Exactly how Cort's process differed from the processes that had been developed by earlier ironmasters along the same lines is not known, but his two inventions together had a tremendous effect on the iron-making industry in Britain; in the next 20 years British iron production quadrupled. The discovery that his partner had invested stolen funds in the enterprise led to Cort's being deprived of his patents and forced into bankruptcy, though he was eventually granted a modest pension.[40]

Another important development was that of metalworking machines by John Wilkinson (1728–1808). In 1755 John Wilkinson became involved with the Bersham concern, a large ironworks at Bersham in North Wales that were known for their high-quality casting. It was also a producer of guns and cannon. It was here that in 1774 Wilkinson patented a technique for a water-powered machine, boring iron guns from a solid piece, rotating the gun barrel rather than the boring bar. This technique made the guns more accurate and less likely to explode. The patent was quashed in 1779 (the navy saw it as a monopoly and sought to overthrow it), but Wilkinson still remained a major manufacturer.

The discovery of external cooling

The period of the second generation of steam engines is the period in which low-pressure steam engines were developed based on the atmosphere vacuum principle. The machines that were developed now tackled the problem of the high energy consumption, due to the repeated heating and cooling of the cylinder in Newcomen's machine. One of the inventors working on improving the low efficiency of the Newcomen engine was James Watt (1736–1819).

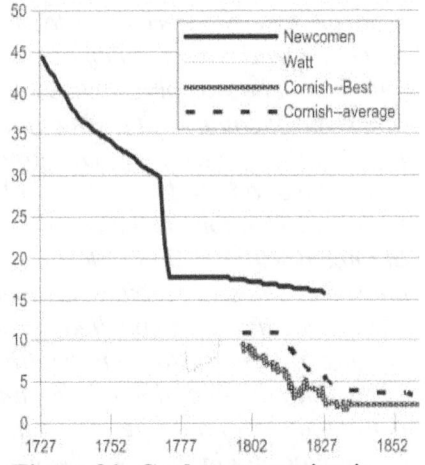

Figure 21: Coal consumption in pumping engines (lbs per HP-hour).
Source: (Allen, 2012) Graph on page 22.

---

[40] Source: *Encyclopedia Britannica*, http://www.britannica.com/EBchecked/topic/138813/Henry-Cort. See also: www.henrycort.net.

Watt's "external condenser" patent immediately doubled steam-engine efficiency. By 1784, Watt's engines were four times more efficient than the old Newcomen engines.

This aspect of energy efficiency is quite important, as shown in Figure 21. Even though Newcomen machines improved in their energy consumption over time, it was Watt who developed a much more energy-efficient concept with his external condenser. The same goes, as we will see further on, for Trevithick's high-pressure steam engine, known as the Cornish machine (Allen, 2012, pp. 17-23).

James Watt's steam engine (1769)

James Watt (1736–1819) was born in Greenock, a Scotch fishing village and seaport on the Firth of Clyde. His father was in the shipping business and served as the town's chief baillie (chief of police), while his mother, Agnes Muirhead, came from a distinguished family and was well educated.

> *When finally sent to the village school, his ill health prevented his making rapid progress; and it was only when thirteen or fourteen years of age that he began to show that he was capable of taking the lead in his class, and to exhibit his ability in the study, particularly, of mathematics...At the age of eighteen, Watt was sent to Glasgow, there to reside with his mother's relatives, and to learn the trade of a mathematical-instrument maker. The mechanic with whom he was placed was soon found too indolent, or was otherwise incapable of giving much aid in the project, and Dr. Dick, of the University of Glasgow, with whom Watt became acquainted, advised him to go to London. Accordingly, he set out in June, 1755, for the metropolis, where, on his arrival, he arranged with Mr. John Morgan, in Cornhill, to work a year at his chosen business, receiving as compensation 20 guineas. At the end of the year he was compelled, by serious ill-health, to return home. Having become restored to health, he went again to Glasgow in 1756, with the intention of pursuing his calling there. But, not being the son of a burgess, and not having served his apprenticeship in the town, he was forbidden by the guilds, or trades-unions, to open a shop in Glasgow. Dr. Dick came to his aid, and employed him to repair some apparatus which had been bequeathed to the college. He was finally allowed the use of three rooms in the University building, its authorities not being under the municipal rule. He remained here until 1760, when, the trades no longer objecting, he took a shop in the city; and in 1761 moved*

again, into a shop on the north side of the Irongate, where he earned a scanty living without molestation, and still kept up his connection with the college. He did some work as a civil engineer in the neighborhood of Glasgow, but soon gave up all other employment, and devoted himself entirely to mechanics (Thurston, 1878, pp. 81-82).

So we find Watt in Glasgow in 1756, where he is a trained instrument maker. His University of Glasgow acquaintances learned of his return and gave him some work. They arranged for permission to set up a shop for Watt on university grounds and created the position "Mathematical Instrument Maker to the University." One day in 1763, Professor John Anderson brought Watt a new problem.

Figure 22: Watt's garret workshop.
Source: On display at the Science Museum, London

The University of Glasgow had a lab-scale model of the Newcomen pump to investigate why the full-scale pumps required so much steam (Figure 23) The model suffered a problem. It would stall after a few strokes. Could Watt repair it? (Scherer, 1965, p. 166)

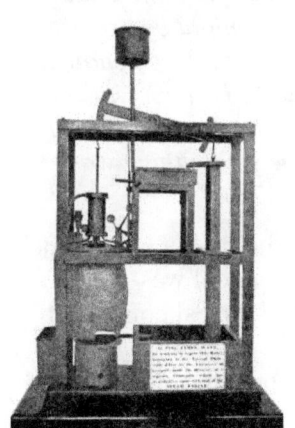

Figure 23: The model of Newcomen's steam engine.

Source: Wikipedia Commons, University of Glasgow

He spent much of his leisure time of which he had, at first, more than was desirable in making philosophical experiments and in the manufacture of musical instruments, in making himself familiar with the sciences, and in devising improvements in the construction of organs. In order to pursue his researches more satisfactorily, he studied German and Italian, and read Smith's "Harmonics," that he might become familiar with the principles of construction of musical instruments. His reading was still very desultory; but the introduction of the Newcomen engine in the neighborhood of Glasgow, and the presence of a model in the college collections which was placed in his hands, in 1763, for repair, led him to study the history of the steam-engine, and to conduct for himself an experimental research into the properties of steam, with a set of improvised apparatus (Thurston, 1878, p. 20).

Soon, as James got in touch with the developments going on around those steam engines, he became fascinated by it. He realized that the economic use of the steam was more important than the mechanical improvement of the engine. He thus arrived at the conclusion that to make the best use of the steam, the cylinder should be kept as hot as the steam itself. The cooling of the steam should be realized outside the cylinder, in a separate vessel. So he developed a separate condenser outside the piston-driven cylinder. He may have been thinking a long time about the problem, but the solution came in a flash:

> *It was in the Green of Glasgow. I had gone to take a walk on a fine Sabbath afternoon. I had entered the Green by the gate at the foot of Charlotte Street—had passed the old washing-house. I was thinking upon the engine at the time, and had gone as far as the Herd's-house, when the idea came into my mind, that as steam was an elastic body it would rush into a vacuum, and if a communication was made between the cylinder and an exhausted vessel, it would rush into it, and might be there condensed without cooling the cylinder. I then saw that I must get quit of the condensed steam and injection water, if I used a jet as in Newcomen's engine. Two ways of doing this occurred to me. First, the water might be run off by a descending pipe, if an offlet could be got at the depth of 35 or 36 feet, and any air might be extracted by a small pump; the second was to make the pump large enough to extract both water and air."* He continued, *"I had not walked farther than the Golf-house [about the site of the Humane Society-house, or a little to the N.W. of that] when the whole thing was arranged in my mind* (Smiles, 1865, pp. 127-128).[41]

To find out whether his idea worked in practice, he built a little model that worked very satisfactorily (Figure 24). A larger model was constructed immediately afterwards, and the result of its test confirmed fully the anticipations that had been awakened by the first experiment. This now proven idea was what changed the development of the atmospheric steam engine, which had run into a dead end, into the

---

[41] This quote can also be found at: Robert Hart, Esq.: *Reminiscences of James Watt*. Meeting of the Society held at Glasgow, on 2 November 1857. Transactions of the Glasgow Archeological Society. Source: http://himedo.net/TheHopkinThomasProject/ TimeLine/Wales/ Steam/URochesterCollection/Hart/Rem.htm.

development of the condenser-based steam engine. It was fundamental, but Watt himself considered his invention a logical step in the range of development around the steam engine:

> *When analyzed, the invention would not appear so great as it seemed to be. In the state in which I found the steam-engine, it was no great effort of mind to observe that the quantity of fuel necessary to make it work would forever prevent its extensive utility. The next step in my progress was equally easy to inquire what was the cause of the great consumption of fuel. This, too, was readily suggested, viz., the waste of fuel which was necessary to bring the whole cylinder, piston, and adjacent parts from the coldness of water to the heat of steam, no fewer than from 15 to 20 times in a minute* (Smiles, 1865, p. 128).

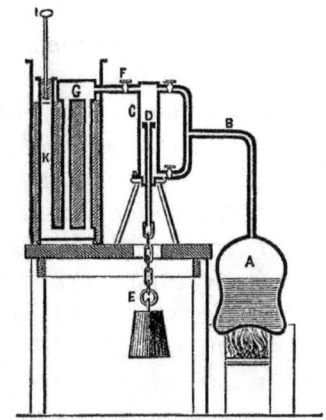

Figure 24: Watt's experimental apparatus (the little prototype of 1765).
Source: Wikimedia Commons

James Watt and his partners in business

To build the larger experimental model, Watt was starting to devote all his time to the project at hand. Then came the classical problem every inventor is faced with: money. Watt's university friends (Black, Anderson) introduced him to John Roebuck (1718–1794), an industrialist who held leases on coal deposits. Roebuck agreed to back the development of a full-scale engine after he saw the model work. Over the next four years, Watt was consumed with making an engine work.

> *When he had concluded to build the larger experimental engine, Watt determined to give his whole time and attention to the work, and hired a room in an old deserted pottery near the Broomielaw. Here he worked with a mechanic—Folm Gardiner, whom he had taken into his employ uninterruptedly for many weeks. Meantime, through his friend Dr. Black, probably, he had made the acquaintance of Dr. Roebuck, a wealthy physician, who had, with other Scotch capitalists, just founded the celebrated Carron Iron-Works, and had opened a correspondence with him, in which he kept that gentleman informed of the progress of his work on the new engine…In August, 1765, he tried the small engine, and wrote*

> *Dr. Roebuck that he had had "good success," although the machine was very imperfect...He then tells his correspondent that he was about to make the larger model. In October, 1765, he finished the latter...Watt was now reduced to poverty, and, after borrowing considerable sums from friends, he was finally compelled to give up his scheme for the time, and to seek employment in order to provide for his family. During an interval of about two years he supported himself by surveying, and by the work of exploring coal-fields in the neighborhood of Glasgow for the magistrates of the city. He did not, however, entirely give up his invention. In 1767, Dr. Roebuck assumed Watt's liabilities to the amount of £1,000[42], and agreed to provide capital for the prosecution of his experiments and to introduce his invention; and, on the other hand, Watt agreed to surrender to Dr. Roebuck two-thirds of the patent. Another engine was next built, having a steam-cylinder seven or eight inches in diameter, which was finished in 1768. This worked sufficiently well to induce the partners to ask for a patent, and the specifications and drawings were completed and presented in 1769* (Thurston, 1878, p. 92).

The contacts between Roebuck and Watt were not by accident. Roebuck, an educated man and well-to-do physician, was a consulting chemist and developed several improvements in processes for the production of chemicals. He went into iron making and had in 1760 established the Carron Company Ironworks at Carron. For that enterprise he needed coal. So he leased from the Duke of Hamilton large coal mines and saltworks at Borrowstounness to supply coal to the Carron works. There he was faced with a massive water problem in the shafts. The Newcomen machines could not cope with that problem, and having heard about Watt and his activities, he contacted Watt. His steam engine was still in its infancy, but he was convinced of its future possibilities and decided to participate. But then fortune changed for Roebuck.

> *Through the expense and loss thus incurred Roebuck became involved in serious pecuniary embarrassments. To his loss by his mines was added that from an unsuccessful attempt to manufacture soda from salt. After sinking in the coal and salt works at Borrowstounness his own fortune, that brought him by his wife, the profits of his other enterprises, and large sums borrowed from friends, he had to withdraw his capital from the Carron ironworks, from the refining works at Birmingham, and*

---

[42] In 2010 this would be worth £1,470,000 using average earnings.

the vitriol works at Prestonpans to satisfy the claims of his creditors. Among Roebuck's debts was one of 1,200 pounds to Boulton, afterwards Watt's well-known partner. Rather than claim against the estate Boulton offered to cancel the debt in return for the transfer to him of Roebuck's two-thirds share in Watt's steam-engine, of which so little was then thought that Roebuck's creditors did not value it as contributing a farthing to his assets (Smiles, 1865, pp. 196-198).

Roebuck went into bankruptcy. In the meantime Watt acquired for this condenser a British patent Nr. 913 in 1769. It would be one of the two important patents that were soon to be challenged. They were also to become a considerable source of revenue that made Watt (and also his later partner Boulton) wealthy.[43] The money he had received from his friend Dr. Roebuck, in today's terms, would be called "angel funding," and it enabled him to continue his work. The consequence was that he had to give up two-thirds of the rights of the patent. So, within this context, came the next set of important moments in his life:

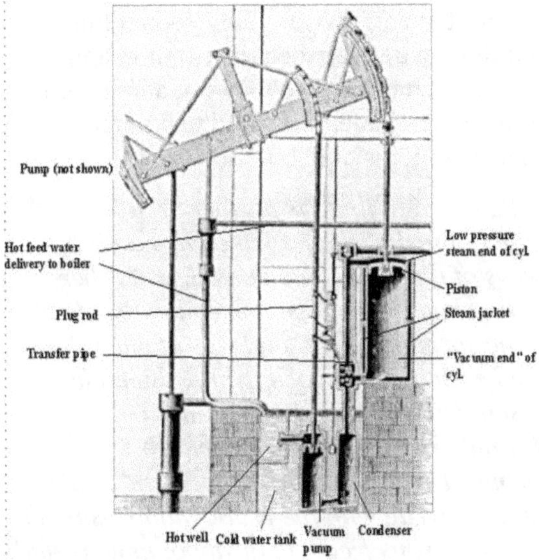

Misfortunes never come singly; and Watt was borne down by the greatest of all misfortunes—the loss of a faithful and affectionate wife while still unable to see a successful issue of his schemes…Watt met Mr. Boulton, who next became his partner, in 1768, on his journey to London to procure his patent, and the latter had then examined Watt's designs, and, at

Figure 25: Watt's steam engine (1774).
Source: Wikimedia Commons, Robert H Thurston: *A history of the growth of the steam engine*

---

[43] For the use of their patent, they charged a license fee that was based on the energy savings the owner of the steam engine would realize. The firm received a third part in value of any saving in fuel for each engine made to their specifications up to 1800. To calculate these savings, Watt developed in 1781 a calculating device that registered the strokes the engine made: Watt's engine counter.

> *once perceiving their value, proposed to purchase an interest. Watt was then unable to reply definitely to Boulton's proposition, pending his business arrangements with Dr. Roebuck; but, with Roebuck's consent, afterwards proposed that Boulton should take a one-third interest with himself and partner, paying Roebuck therefore one-half of all expenses previously incurred, and whatever he should choose to add to compensate "for the risk he had run." Subsequently, Dr. Roebuck proposed to transfer to Boulton and to Dr. Small, who was desirous of taking interest with Boulton, one-half of his proprietorship in Watt's inventions, on receiving "a sum not less than one thousand pounds," which should, after the experiments on the engine were completed, be deemed "just and reasonable"* (Thurston, 1878, p. 92).

Boulton, the son of a wealthy silver merchant, having a factory in Soho, two miles distant from Birmingham, had been doing some work on a steam engine himself. So he rapidly understood the relevance of the work done by Watt. So in March 1773, Boulton acquired Roebuck's rights to the engine, four years after the engine was patented and nine years after Watt first discovered the separate condenser. Boulton and Watt's personalities complemented each other, and they got along well. Boulton's assembly of accomplished craftsmen provided the much-needed expertise that Watt had lacked in his collaboration with Roebuck.

> *Before Watt could leave Scotland to join his partner at Soho, it was necessary that he should finish the work which he had in hand, including the surveys of the Caledonian canal, and other smaller works, which he had had in progress some months. He reached Birmingham in the spring of 1774, and was at once domiciled at Soho, where he set at work upon the partly-made engines which had been sent from Scotland some time previously...It was in November, 1774, that Watt finally announced to his old partner, Dr. Roebuck, the successful trial of the Kilmeil engine. He did not write with the usual enthusiasm and extravagance of the inventor, for his frequent disappointments and prolonged suspense had very thoroughly extinguished his vivacity. He simply wrote: "The fire engine I have invented is now going, and answers much better than any other that has yet been made; and I expect that the invention will be very beneficial to me"*
> (Thurston, 1878, p. 97).

Boulton was right: both the technical characteristics of Newcomen's steam engine and the license cost involved sparked a high interest in

Watt's machine by mine owners and the municipality of London, which needed to pump water to the inhabitants of London. He wrote to Watt:

> *The people in London, Manchester and Birmingham are steam mill mad. I don't mean to hurry you, but I think in the course of a month or two, we should determine to take out a patent for certain methods of producing rotative motion...There is no other Cornwall to be found, and the most likely line for the consumption of our engines is the application of them to mills which is certainly an extensive field* (letter from Boulton to Watt, 21 June 1781).[44]

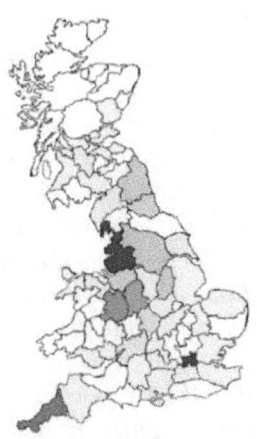

4: Boulton & Watt Engines, 1775-1800

Figure 26: Geographical diffusion of steam technology, Boulton & Watt steam engines (1775–1800).

Legend: Darker areas indicate an increasing number of steam engines erected.
Source: (A. Nuvolari et al., 2003)

Steam Engine Act (1775)

So the company was going to grow, but not without problems of its own. One of them was that the patent protection was going to end in 1783. So Watt, in 1774, applied for an extension of his patent right, and he obtained the extension being granted for twenty-four years, as decided in the Steam Engine Act[45] in 1775. The combination of the patent and the act made Watt invincible in the realms of steam engineering at that time and gave him such an overall monopoly that it would have seemed foolish to challenge it:

> *The firm was therefore at once driven to make preparations for a large business. The first and most important matter, however, was to secure an extension of the patent, which was soon to expire. If not renewed, the 15 years of study and toil, of poverty and anxiety, through which Watt had toiled, would prove profitless to the*

---

[44] Ann Sproule, *James Watt*, Exley Publications, Herts, United Kingdom, 1992. See also: Watt Biography. Source: http://www.egr.msu.edu/~lira/supp/steam/wattbio.html.
[45] 1775 Steam Engines Act: "An Act for vesting in James Watt, Engineer...the sole Use and Property of certain Steam Engines...of his Invention...throughout his Majesty's Dominions for a limited time."

inventor, and the fruits of his genius would have become the unearned property of others. Watt saw, at one time, little hope of securing the necessary act of Parliament, and was greatly tempted to accept a position tendered him by the Russian Government, upon the solicitation of his old friend, Mr. Robinson, then a Professor of Mathematics at the Naval School at Cronstadt. The salary was £1,000 a princely income for a man in Watt's circumstances, and a peculiar temptation to the needy mechanic. Watt, however, went to London, and, with the help of his own and of Boulton's influential friends, succeeded in getting his bill [Act of Parliament] through. His patent was extended 24 years, and Boulton & Watt set about the work of introducing their engines with the industry and enterprise which characterized their every act. In the new firm, Boulton took charge of the general business, and Watt superintended the design, construction, and erection of their engines. Boulton's business capacity, with Watt's wonderful mechanical ability—Boulton's physical health, and his vigor and courage, offsetting Watt's feeble health and depression of spirits and, more than all, Boulton's pecuniary resources, both in his own purse and in those of his friends, enabled the firm to conquer all difficulties, whether in finance, in litigation, or in engineering (Thurston, 1878, p. 103).

Watt and Boulton soon became formal partners: Boulton for two-thirds and Watt for one-third of the revenues of the patents. His financial position now secured, Watt was relieved of the uncertainties regarding his business connections, and he married a second wife.

Perfecting the engine

Watt continued perfecting his steam engine, acquiring in the period 1775 to 1785 five patents, covering a large number of valuable improvements upon the steam engine, as well as several independent inventions. In the later years, more patents were added (Table 1). As Watt's steam engines were originally used creating a unidirectional movement (as in vertical), the first applications were found in pumping (e.g., pumping water from the mines, drinking water to villages). However, a rotary motion (as in a wheel) was interesting. This would open a range of new applications.

Watt invented his original wheel-engine, but that was not practically used. He prepared a model in which he made use of a crank connected with the working beam of the engine so as to produce the necessary rotary motion. The crank principle was not new, as it was one of the most common of mechanical appliances. It was in daily use in every spinning wheel and in every turner's and knife-grinder's foot-lathe. However,

Table 1: Overview of James Watt's later patents

Year and number (granted)	Description/Application
British Patent 1306, October 25, 1781	Certain new methods of applying the vibrating or reciprocating motion of steam or fire engines, to produce a continued rotative or circular motion round an axis or center, and thereby to give motion to the wheels of mills or other machines.
British Patent 1321, March 12, 1782	Description of Expansive use of Steam, Double-Acting Engines and Compounding, Rotative Engines. Certain new improvements upon steam or fire engines for raising water, and other mechanical purposes, and certain new pieces or mechanism applicable to the same.
British Patent 1432, April 28, 1782	Various mechanisms including the Parallel Motion, the Balance of Pumping Rods, Steam Hammers, General Application of Steam Power in Mills etc., and the Application of Steam power to Carriages etc.
British Patent No. 1485, June 14, 1785	Smokeless Furnaces & Fire-places

Source: http://himedo.net/TheHopkinThomasProject/TimeLine/Wales/Steam/JamesWatt/RobinsonMusson/JamesWattPatents.htm#Patent1781

James Wasbrough had already patented the crank and wheel principle (British Patent 1213, 10 March 1779). Also, James Pickard of Birmingham had in 1780 patented a counterweighted crank device (British Patent 1263, 23 August 1780). To circumvent these patents, Watt had therefore to employ some other method. He adapted, after investigation, five different concepts for securing rotary motion without a crank: the "inclined wheel," "counterweighted crank wheel," "eccentric wheel," "eccentric wheel with internal driving wheel," and "sun-and-planet wheel," The invention was patented in February 1782 (together with patents for the other methods).

There was an interesting development related to the license structure that was applied.

> *Many of the contracts of Boulton & Watt gave them, as compensation for their engines, a fraction usually one-third of the value of the fuel saved by the use of the Watt engine in place of the engine of Newcomen, the amount due being paid annually or semi-annually, with an option of redemption on the part of the purchaser at ten years' purchase* (Thurston, 1878, p. 114).

This form of agreement compelled a careful determination, often, of the work done and fuel consumed by both the engine taken out and that put in its place. The problem was how to calculate those savings reliably. This was important for both Watt and the mineowner as it determined what he had to pay. So Watt developed a device to calculate the savings, the engine counter (Figure 27).

Figure 27: Watt's engine counter, c. 1781.

Source: Science Museum/Science & Society Picture Library, London

Watt next began to develop a series of minor inventions, such as the governor and the mercury steam-gauge, a barometer to measure the pressure of the steam. These inventions were secured by his patent of 28 April 1782, together with the steam tilt-hammer and a steam-carriage or "locomotive engine." This was the first application of the steam engine in a not-fixed situation, as it was supposed to be mobile by nature.

Watt's steam engine was a great success, especially in Cornwall, for between 1777 and 1801, some fifty-two engines were erected. The typical agreement that Boulton & Watt stipulated with Cornish mine entrepreneurs (commonly termed "adventurers") was that the two partners would provide the drawings and supervise the construction of the engine. They would also supply some particularly important components of the engine, such as some of the valves. During the time that Boulton & Watt were operating in Cornwall, they netted a total of £180,000 pounds (equivalent to £245,000,000 in 2010 using average earnings) in royalty payments for their work (Kelly, 2002).

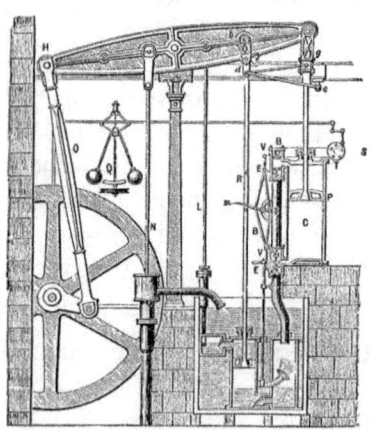

Figure 28: Sketch showing a rotative steam engine designed by Boulton & Watt, England, 1784.

Source: Wikimedia Commons. Robert H Thurston: *A history of the growth of the steam engine*

Contemporary developments

The steam engines that were based on Watt's design, despite their undoubtedly great advance in efficiency compared to Newcomen's engines, were nonetheless thermodynamically quite inefficient. Above that was the fact that the creation of a vacuum remained essential to the operation of this type of engine. This all was restricting the performance, efficiency, and flexibility of the steam machines. So inventive individuals were looking to improve the steam engines—individuals such as Edward Bull (1759–1798), Jonathan Hornblower (1753–1815), William Symington (1764–1831) and Richard Trevithick (1771–1833).

How were the steam engines built?

One could think that the steam engines developed by Newcomen, Symington, Watt, and so on were all built by one manufacturer. That was not the case, as they were more of a combined engineering effort than a single manufacturer's product:

> *Steam engines were rarely built by one organization or individual in the 18th century; more often they were the product of several. This content downloaded from ironworks, carpenters and stonemasons, and an engineer or erector. The cylinder might come from one foundry, the rest of the ironwork from another. The timberwork and stonework for the framing and engine house would be done by local craftsmen or by the employees of the purchaser, and the erection of the parts to form a working engine would be supervised by a professional erector who was in effect a consulting mechanical engineer. The boilers, moreover, might come from yet a further different source. Before the opening of the Soho foundry in 1796, for example, Boulton and Watt merely supplied the plans, some of the more complicated metal parts such as valves and nozzles, and the expertise to erect the machinery and make it work. They did, however, act as middlemen for the supply of boilers, cylinders, and so on, and usually insisted that the customer pay them and not the ironworks who was actually making the parts. The customer was responsible for obtaining the other parts locally, a situation which sometimes caused misunderstandings between the firm and its clients* (Kanefsky & Robey, 1980, p. 173).

These engines were built by "engineers," often people from a humble (that is to say, non-aristocratic) background, that had hands-on experience and could translate the conceptual ideas of the "gentlemen of science" and the prototypes of the "inventors" into working and performing products.

Bull's steam engine

Edward Bull (1759–1798), formerly an erector of Watt's machines and working closely together with Richard Trevithick, designed a steam engine that was then known as the "Bull Cornish Engine." It was used in the Ding Dong mines in Cornwall. His work was closely related with that of Jonathan Hornblower.

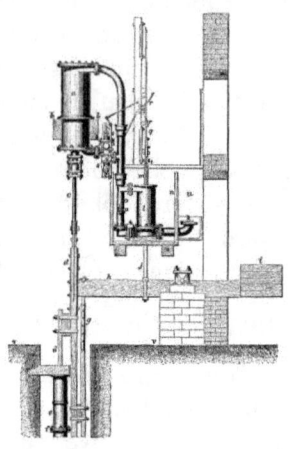

Figure 29: Bull Cornish Engine (1798).
Source: Robert H. Thurston 1878"
A history of the growth of the steam engine

> *The operation of the engine is similar to that of a Watt engine. It is still in use, with a few modifications and improvements, and is a very economical and durable machine. It has not been as generally adopted, however, as it would probably have been had not the legal proscription of Watt's patents so seriously interfered with its introduction. Its simplicity and lightness are decided advantages, and its designers are entitled to great credit for their boldness and ingenuity, as displayed in their application of the minor devices which distinguish the engine. The design is probably to be credited to Bull originally; but Trevithick built some of these engines, and is supposed to have greatly improved them while working with Edward Bull, the son of the inventor, William Bull* (Thurston, 1878, p. 140).

Hornblower's compound steam engine

Jonathan Hornblower Sr (1753–1815) came from an engineering family. His grandfather, Joseph Hornblower, had worked for Joseph Newcomen. In 1778 Jonathan was working at the Ting Tang mine in Cornwall, where he was helping Watt & Boulton to erect one of their steam engines. His son, Jonathan Carter Hornblower, in 1781 patented a two-cylinder "compound" engine. In this engine the steam pushes on one piston (as opposed to pulling via vacuum, as in previous designs), and when it reaches the end of its stroke is transferred into a second cylinder that exhausts into a condenser as "normal." The text of the patent was as follows:

NOW KNOW YE, that, in compliance with the said proviso, and in pursuance of the said statute, I, the said Jonathan Hornblower, do hereby declare, that my said invention is described in manner and form following: that is to say, first, I use two vessels in which the steam is to act, and which, in other steam engines, are generally called cylinders. Secondly, I employ the steam, after it has acted in the first vessel, to operate a second time in the other, by permitting it to expand itself, which I do by connecting the vessels together, and forming proper channels and apertures, whereby the steam shall occasionally go in and out of the said vessels. Thirdly, I condense the steam, by causing it to pass in contact with metalline surfaces, while water is applied to the opposite side. Fourthly, to discharge the engine of the water used to condense the steam, I suspend a column of water in a tube or vessel constructed for that purpose on the principles of the barometer; the upper end having open communication with the steam vessels, and the lower end being immersed into a vessel of water. Fifthly, to discharge the air which enters the steam vessels with the condensing water, or otherwise, I introduce it into a separate vessel, whence it is protruded by the admission of steam. Sixthly, that the condensed vapour shall not remain in the steam vessel in which the steam is condensed, I collect it into another vessel, which has open communication with the steam vessels, and the water in the mine, reservoir, or river. Lastly, in cases where the atmosphere is to be employed to act on the piston, I use a piston so constructed as to admit steam round its periphery, and in contact with the sides of the steam vessel, thereby to prevent the external air from passing in between the piston and the sides of the steam vessel.[46]

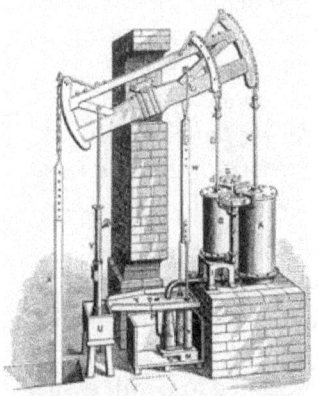

Figure 30: Jonathan Hornblower's compound engine (1781).

Source: Science Museum/Science & Society Picture Library, London

Hornblower's design was more efficient than Watt's single-acting designs, but similar enough to his double-acting system that Boulton and

---

[46] British Patent No. 1298 of 1781.

Watt were able to have the patent overturned by the courts in 1799, an action that took quite some effort and resulted in something like a "patent war."

Patent war

Elements of Watt's design, especially the separate condenser, were incorporated in many "pirate" engines. Watt had to protect his patent position, for example in the case of the machines made by Jonathan Hornblower Jr, and also against the numerous "Cornish pirates" who infringed on his patent. But going to court on a patent infringement certainly has its risks.

> *In one type of case the defendant had simply manufactured reproductions of Watt's engine, knowing full well that he was infringing. The defendant may have made just a few machines for his own use or for sale. The Manchester firm of ironmongers Bateman and Sherratt was an example of this kind of infringement, and John Wilkinson, who had made parts for Watt's engines, was another. Another type of piracy or alleged piracy came from patentees of other engines. An inventor might use a part for a new invention claimed by Watt to be by his own patent. Drawings and descriptions for new patents were made public, thus giving holders of earlier patents a chance to examine them for a possible infringement. Watt believed that Jonathan's patent of 1781 infringed his 1769 patent, but after considering the cost and likelihood of losing at trial, he held off, while waiting to see how successful Jonathan would be* (Howard, 2009, p. 11).

In 1781 Jonathon Hornblower Jr. was granted his patent (British Patent №. 1298 of 1781) for a double-cylinder steam engine. This patent precluded Watt from using the expansion of steam in a second cylinder of greater diameter than the first. The extension of the patent failed, and it expired in 1795. And there was another patent granted to a mill owner named James Pickard (British Patent № 1263, 23 August 1780) in 1780 to cover the crank. Then, in 1782, Jonathan Hornblower erected a compound engine at the Radstock Colliery near Bristol, for a Mr. Winwood.

Hearing of this, Watt put an advertisement in the Bristol papers, where Boulton & Watt threatened legal action against him, along with all other makers of engines featuring a "Piston pressed down by steam." Jonathan's brother, Jabez Hornblower, called Watt's bluff by an article called "History of the Steam Engine" in Gregory's *Mechanics* and in his

"Address to the Cornish Miner," written in May 1788 together with John Winwood.[47]

Jabez Hornblower and James Watt were about the same age. Jabez had worked erecting Watt engines along with his father and younger brothers. Over time their relation changed. In 1780 Jabez settled in Bristol; however, his business failed in 1786, and he was declared bankrupt. By 1790 Jabez had moved to London and was established as a maker of steam engines and other machinery. He went into business with J. A. Maberly, a London currier (Howard, 2009, p. 11). Meanwhile Watt was still looking to protect his patent.

> *As much as Boulton and Watt might wish to avoid an expensive lawsuit, they believed that a failure to act against one infringer would signal others to ignore the patent, resulting in loss of future revenues. In 1793 the business partners brought an action against former employee Edward Bull, who in 1792 began building engines in Cornwall that Watt claimed were essentially the same as his. The case began in the Court of Common Pleas before Lord Chief Justice Eyre and a special jury. Bull defended himself by calling into question the validity of Watt's patent on the basis of the insufficiency of the specifications. While the court found for the plaintiffs on the infringement, the judges were evenly divided for and against the validity of the patent. The case dragged on unsettled until 1799. Boulton and Watt clients in Cornwall watched the case closely. Taking a chance that Watt would lose as had Arkwright, the Cornish adventurers began defaulting on payments* (Howard, 2009, pp. 11-12).

So finally, as they were not paying him anymore, Watt decided to pursue the "pirates." The first suit was against engineer Edward Bull in 1792, the second against Jonathan Hornblower and his partner Stephen Maberly for infringement of Watt's patent in 1796: Boulton & Watt v. Bull and Boulton & Watt v. Hornblower & Maberly (1796). It was Jabez Carter Hornblower who was faced with the lawsuit and its trial.

> *On 1 January 1796, Boulton and Watt obtained an injunction in the Court of Chancery against Hornblower and Maberly restraining them from building further engines, and against the owner of the colliery restraining him from using the engine. Maberly then negotiated with Matthew Boulton in an effort to get*

---

[47] "History of the Steam Engine: Historical review of improvements in the steam engine in the XVIII century," by Jabez Carter Hornblower, printed in the first and second editions of O. Gregory's *Mechanics*, Vol. II, pp. 358–360.

> *relief from having to pay premiums on engines already built in return for a promise that he and his partner would give security not to infringe in the future. Watt might have conceded on the royalties, but Hornblower and Maberly defied the injunction and with the help of Arthur Woolf, a Cornish engineer, began building a new engine for the Meaux & Co. brewery in London. (Modern spelling is Meux.) Despite the expense and uncertainty of upholding their patent, Boulton and Watt proceeded to take the legal side of their case to trial in the Court of Common Plea* (Howard, 2009, p. 13).

This all happened because Jabez and Maberly had installed a steam engine near Newcastle in the northeast of England. In the meantime in Cornwall, there were things happening.

> *As each side gathered affidavits and served subpoenas, Watt's son dealt with a growing problem in Cornwall. The adventurers from the Poldice mine wanted abatement on the premiums on one of their engines, but Watt hated to set a precedent that could result in loss and a plague of new problems with other mine owners. Watt Jr. wrote a counter-proposal asking for a lump sum payment for a smaller amount [which was accepted]. Watt Jr. was pleased with the victorious outcome...He told Wilson that Hornblower and Maberly immediately asked to meet with him to work out a compromise, but the son, feeling confident, insisted on the same terms that "other pirates" had accepted: full payment of the premium and a penalty bond not to infringe in the future and all costs incurred by the defendants in the proceedings. He and his father believed that Hornblower and Maberly had financial support "from Cornwall." Watt was anxious to find out who those supporters were and promised that those who paid part of the defendants' expenses would top the list for the next infringement action* (Howard, 2009, p. 14).

Finally on 16 December 1796, the case was tried in the Court of Common Pleas before a special jury.

> *Lord Chief Justice James Eyre, who had presided over the Bull case, presided again over this historic case. Boulton and Watt had called a number of eminent engineers and scientists, Fellows of the Royal Society, to testify, including Watt's old friend from the University of Glasgow, Professor John Robinson. The two dozen witnesses subpoenaed also included John Roebuck, an early backer of Watt's invention, Thomas Wilson, Watt's agent in*

Cornwall, and William Murdock, the engineer who had replaced Jabez at Donnington Wood...The most prominent witness was Joseph Bramah, inventor of the hydraulic press and other useful items. Bramah, whose testimony was cut short at trial, published a lengthy letter in 1797 addressed to the Chief Justice. (Watt called it ninety pages of unorganized nonsense.) The defendants also had the support of many who opposed monopolies and who believed that Watt had deliberately drawn his specifications to be as obscure as possible...The witnesses for Boulton and Watt stated that Jabez and other members of the Hornblower family had worked for Watt and had derived knowledge of the steam engines through their employment. Family members had access to the drawings of the engines. In turn Hornblower and Maberly attacked the patent on the grounds that the specifications were vague and insufficient. Jabez maintained the separate condenser was not an original invention of Watt that could be protected by a patent. He claimed it was an adaptation of an improvement to the Newcomen engine by "a Mr. Gainsborough." This was Humphrey Gainsborough, mentioned above, the brother of artist Thomas Gainsborough, who claimed to have invented the separate condenser before Watt had patented his...The defendants failed to make their case, and Boulton and Watt emerged the victors.

In February 1797 Hornblower and Maberly made a motion for a new trial. The counsel for Watt thought it was a frivolous motion brought as a means to delay payment of costs and to load the trial record with objections. The court rejected the motion. The next step was to lodge an appeal, which the defendants did, allowing them another delay in payments, although following the verdict against him Maberly closed his business and discharged his workers. James Watt Jr. presumed that Maberly and Hornblower had gone to Cornwall to rally their supporters. At the end of December James Watt Jr. wrote to Wilson that friends had said that the engineering business had cost Maberly at least £8,000.[48] *...The case dragged on through 1798 when counsel for both sides argued the issues on two separate occasions "with great ability" according to the opinion of the justices. The decision was announced in 1799, with a verdict for Boulton and Watt on the crucial point of the separate condenser—Watt's 1969 patent was valid and Jabez had infringed it. The appeal also*

---

[48] In 2010, £8,500 from 1779 is worth £12,000,000 using an average earnings-based calculation. Source: www.measuringworth.com.

> *settled matters in the case of Edward Bull, who was also judged to have infringed Watt's 1769 patent. The partners now set out to collect the rest of the delinquent premiums...Boulton and Watt settled with the rest of the Cornish adventurers* (Howard, 2009, pp. 15-16).

Then Boulton and Watt collected their license fees due from the owners of the mill using his steam engine. For Jabez Hornblower the case did not end too well as his next enterprise failed also, and he had debts of thousands of pounds.

> *From 1803 to 1805 Jabez was imprisoned in the King's Bench Prison in Southwark, a debtor's prison...Although considered to be better than other prisons, King's Bench was still dirty, overcrowded and prone to outbreaks of typhus. Prisoners had to provide their own food, drink, and bedding. Eventually his family was able to get £2,000 pounds from his estate to set him free. (Possibly this sum came from his parents' settling of his mother's estate.)...Jabez...died in London on 11 July 1814* (Howard, 2009, pp. 17, 19).

Boulton & Watt: the end of a partnership

By 1790 Watt was a wealthy man, having received £76,000[49] in royalties on his patents in eleven years. Boulton & Watt constructed a total of 496 engines between 1775 and 1800 when the patent expired. Of these 164 were pumping engines, twenty-four were blowing engines for blast furnaces, and 308 for driving machinery; the latter are almost certain to have been rotative. After the expiry of Watt's patent in 1800, it seems that Boulton & Watt concentrated on the rotative engine, for the company made a far greater number of those than of the up-and-down pumping engines (Kelly, 2002). Which were fewer than the about 1,500 Newcomen engines that from 1712 until 1800 had been built in the United Kingdom.

> *The co-partnership of Boulton & Watt terminated by limitation, and with the expiration of the patents under which they had been working, in the first year of the present century; and both partners, now old and feeble, withdrew from active business, leaving their sons to renew the agreement and to carry on the business under the same firm style. Boulton, however, still interested himself in*

---

[49] In 2010 this would have been worth £93 million using average earnings as a base for the calculation.

some branches of manufacture, especially in his mint, where he had coined many years and for several nations (Thurston, 1878, pp. 126-127).

This was also the moment for Watt to retire.

Watt retired, a little later, to Heathfield, where he passed the remainder of his life in peaceful enjoyment of the society of his friends, in studies of all current matters of interest in science, as well as in engineering. One by one his old friends died: Black in 1799, Priestley, an exile to America, in 1803, and Robinson a little later. Boulton died, at the age of eighty-one, August 17, 1809, and even the loss of this nearest and dearest of his friends outside the family was a less severe blow than that of his son Gregory, who died in 1804....He died August 19, 1819, in the eighty-third year of his age, and was buried in Handsworth Church. The sculptor Chantrey was employed to place a fitting monument above his grave, and the nation erected a statue of the great man in Westminster Abbey (Thurston, 1878, p. 128).

Applications of the steam engine

As indicated before, the linear-movement steam engines were used in applications where no rotative movement was needed. The pumping of water out of mines was such an application. But there were numerous other applications. Watt's steam engine was used in paper mills, flour mills, cotton mills, iron mills, distilleries, and canals and waterworks. They had one thing in common: they were stationary applications. There were others who paid attention to the mobile application of the steam engine.

William Symington's steam engine

William Symington (1764–1831) was educated to be a minister at the University of Glasgow and the University of Dublin. The ministry had slight attraction for him, though, and when the time came for him to choose a profession, he adopted that of civil engineering. He developed a steam engine, combining the efficiency of the Watt engine with the simplicity of the steam engine devised by Thomas Newcomen. He patented his idea in 1787. Symington was quick to understand the other applications for the steam engine besides pumping mines. He worked on the application in carriages and boats. As the road application proved impractical, he moved successfully to application of steam engines in boats and became a pioneer of steam shipping.

In 1786 he worked out a model for a steam road-car. This was regarded very highly by all who saw it. It is said that Mr. Meason, manager of the lead mines at Warlockhead, was so pleased with the model, the merit of which principally belonged to young Symington, that he sent him into Edinburgh for the purpose of exhibiting it before the professors of the University, and other scientific gentlemen of the city, in the hope that it might lead in some way to his future advancement in life. Mr. Meason became the patron and friend of Symington, allowed the model to be exhibited at his own house, and invited many persons of distinction to inspect it. The carriage supported on four wheels had a locomotive behind, the front wheels being arranged with steering-gear. A cylindrical boiler was used for generating steam, which communicated by a steam-pipe with the two horizontal cylinders, one on each side of the firebox of the boiler. When steam was turned into the cylinder, the piston made an outward stroke; a vacuum was then formed, the steam being condensed in a cold water tank placed beneath the cylinders, and the piston was forced back by the pressure of the atmosphere. The piston rods communicated their motion to the driving-axle and wheels through rack rods, which worked toothed wheels placed on the hind axle on both sides of the engine, and the alternate action of the rack rods upon the tooth and ratchet wheels, with which the drums were provided, produced the rotary motion. The boiler was fitted with a lever and weight safety valve. Symington's locomotive was abandoned, the inventor considering that the scheme of steam travel on the common roads was impracticable (Weeks, 1904, p. 46).

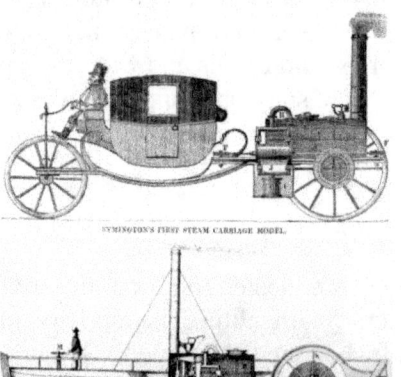

Figure 31: Symington's designs for a steam carriage (1786) and a steam tug (1803)

Source: top (Rankine & Rankine, 1862); bottom Wikimedia Commons, drawing by Bowie, 1883.

By 1800 Symington had started working on a horizontal steam engine. In 1801 he got a patent for his design. Symington died in 1831.

A cluster of innovations

Starting from the design of Newcomen's machine, James Watt improved upon the Newcomen concept by adding his external condenser. An improved fuel efficiency was the key result. The idea might have been logical to him, but realizing the first prototypes needed more than only technical skills. Seed-financing the early work through people in his environment, he was so lucky as to partner up with Boulton, a man with a character and experience quite complementary to Watt's. It would be the start of a fruitful cooperation.

Improving further on the original design, Watt continued making his machine technically and economically better. Protecting his ideas with patents, his single-acting machine grew into the double-acting machine, using the "supporting" inventions created by Cort and Wilkinson. This way Watt could improve his cylinders considerably. But there was also the other side of the business, where the inventors had to protect their intellectual property. Watt & Boulton, after they managed to secure the patent by its extension in 1775, had to address the infringements by the "pirates" who took their design and built similar engines without paying royalties. With the expiration of the patent in sight (1800), they started

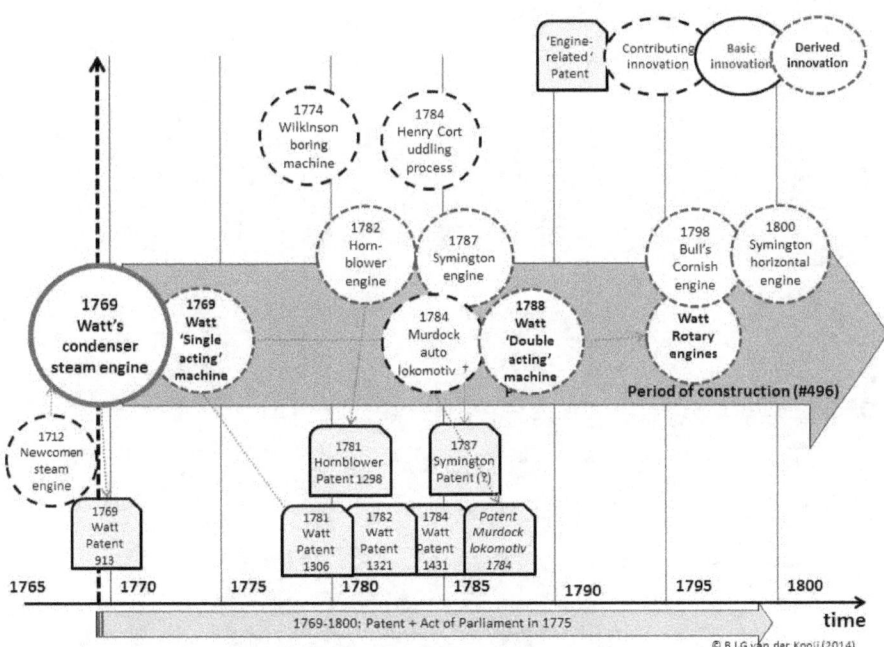

Figure 32: Cluster of innovations around Watt's engine.

#496: Over time, up to the early 1800s, there would be constructed about 500 engines.
Source: Figure created by author

after quite some deliberation in the 1790s with infringement suits. They succeeded in uphold their rights and received their payments due.

All in all, one can conclude that the contributions Watt made to the further development of the steam engine were impressive and had a great technical impact. It certainly solved the water problem of the mines. But next to that, he was fulfilling the "power needs" of the Industrial Revolution, and that impact would be even greater. It was his basic idea of the condenser that resulted in a revolutionary design of the seam engine (Figure 32).

Third-Generation Steam Engines (1800+)

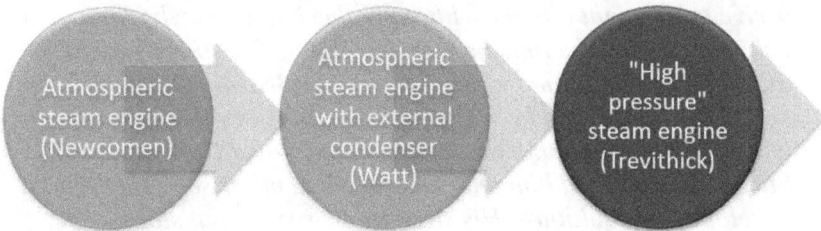

In the third generation of steam engines, high-pressure steam was to replace the atmospheric vacuum principle. As the external condenser principle, heavily protected and anxiously guarded by Watt and Boulton, was mainly successfully applied in stationary applications, the smaller compact high-pressure boilers became the building blocks for the development of mobile steam-driven applications that would have a great impact: the second Industrial Revolution. The often huge stationary machines now suddenly became small and compact and proved, after decades of trials, fortunes, and misfortunes, to be the building blocks of mobile energy.

High pressure eliminates the external condenser

James Watt secured his inventions by covering them with numerous patents, the 1769 condenser-patent being an important one. The patent gave Watt & Boulton protection till 1800, when it expired. It earned Watt & Boulton their license fees, and it also prohibited other manufacturers making similar versions of the steam machine. As Watt's monopoly obstructed the making of those machines, other developments trying to circumvent his "external condenser patent" and its resulting monopoly took place. Among those was Richard Trevithick, who developed a high-pressure steam engine that did not need a condenser.[50]

> *Richard Trevithick's father, Richard Francis (1735–1797), was manager (or captain) at a number of Cornish mines...He earned £2 a month from each...Richard senior was a Methodist class leader and knew one of the movement's founders, the famous preacher John Wesley (1703–91)...Trevithick was born on 13th April 1771...The young Trevithick grew to be tall and strong—and headstrong—and he was indulged by his mother and sisters. He attended school in Camborne where he was taught the "three Rs," though he never mastered spelling. His unconventional approach to arithmetic annoyed his schoolmaster as his quick brain would arrive at the right answer without bothering with orthodox calculations. By the time he was 19 (1790), Trevithick was 1.88m tall (6ft 2in)—at a time when the average British male was 1.7m in height (5ft 7in)—and he was generally known as the Cornish Giant. It is claimed that doctors from the Royal College of Surgeons examined him and said they had never seen such finely developed musculature. His strength was such that stories circulated about him, saying he could lift a 500kg blacksmith's mandrel (the tapered cast iron pipe used for shaping), hurl a sledge hammer over the top of an engine house and write his name on an overhead surface while a 25kg weight was attached to his thumb. He was also a noted Cornish wrestler.*[51]

Richard Trevithick (1771–1833) was a naturally skillful mechanic and was placed by his father with Watt's assistant, Murdoch, who was superintending the erection of pumping engines in Cornwall. Trevithick had also been assisting Edward Bull in erecting low-pressure condensing engines. When in 1796, Boulton & Watt served the partners with an injunction for infringing Watt's condenser patent, Trevithick (giving evidence at the lawsuits) at once turned his efforts towards inventing around Watt's patent by making an engine that did not need a condenser.

---

[50] In a high-pressure steam engine, creating a vacuum is unnecessary, for the expansive force of steam alone is capable in such an engine of working a piston with a force proportional to the steam's pressure. It follows that high-pressure engines don't require condensers, external or otherwise. A condenser could, to be sure, contribute to the fuel efficiency of a high-pressure engine, but it need not and often did not contribute to its overall cost effectiveness.

[51] Text from the biography provided on the website Engineering Timelines (accessed June 2014). Source: http://www.engineering-timelines.com/who/Trevithick_R/TrevithickRichard2.asp.

He was in a unique position to do so, being acquainted with both Bull's and Hornblower's work.

In the years around 1800, he built some steam-powered water pumps, for example, the one at Cook's Kitchen, located close to the Dolcoath, where his father worked in the mine (see Figure 33). Soon, in the period of 1796 to 1799, he developed working models of a high-pressure machine, for example, the Kensington model (see Figure 34).

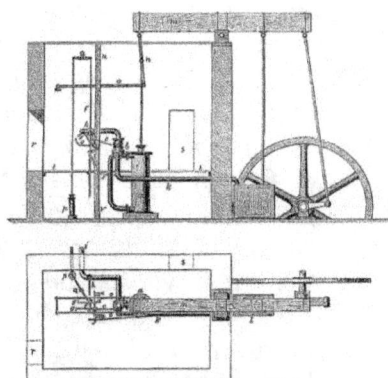

Figure 33: Trevithick's high-pressure expansive steam-condensing whim-engine, erected at Cook's Kitchen (1800).

Source: (Trevithick, 1872, p. 36), www.engineering-timelines.com/who/Trevithick_R/trevithickRichard4A.asp

> *Lord and Lady Dedunstanville, the large landed proprietors in the mining district—embracing Dolcoath, Cook's Kitchen, Stray Park, and many more of the early Cornish mines—and Mr. Davies Gilbert, a friend of Trevithick's, came to the house to see the model work. A boiler, something like a strong iron kettle, was placed on the fire; Davies Gilbert was stoker, and blew the bellows; Lady Dedunstanville was engine-man, and turned the cock for the admission of steam to the first high-pressure steam-engine. The model was made of bright brass* (Trevithick, 1872, p. 103).

The following letter from Gilbert Davies also illustrates the person of Trevithick (Trevithick, 1872, p. 68):

My dear Sir, East Bourne, April 29, 1839

I will give as good an account as I can of Richard Trevithick. His father was the chief manager in Dolcoath Mine, and he bore the reputation of being the best

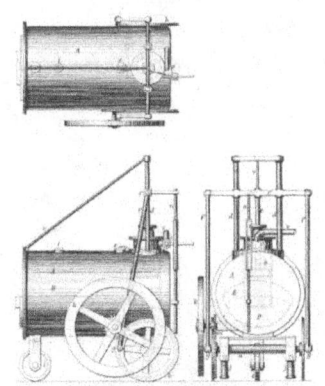

Figure 34: Trevithick's steam engine: the Kensington model (1798).

Source:
http://himedo.net/TheHopkinThomasProject/TimeLine/Wales/LocomotiveDevelopment/Richard%20Trevithick/RichardTrevithick.htm

informed and most skilful captain in all western mines; for as broad a line of distinction was then made between the eastern and western mines (the Gwennap and the Camborne lines) as between those of different nations. I knew the father very well, and about the year 1790 I remember hearing from Sir. Jonathan Hornblower that a tall and strong young man had made his appearance among engineers, and that on more than one occasion he had threatened some people who contradicted him to fling them into the engine-shaft. In the latter part of November of that year I was called to London as a witness in a steam-engine cause between Messrs. Boulton and Watt and Malberly. There I believe that I first saw Richard Trevithick, jun., and certainly there I first became acquainted with him. Our correspondence commenced soon afterwards, and he was very frequently in the habit of calling at Tredrea to ask my opinion on various projects that occurred to his mind—some of them very ingenious, and others so wild as not to rest on any foundation at all. I cannot trace the succession in point of time. On one occasion Trevithick came to me and inquired with great eagerness as to what I apprehended would be loss of power in working an engine by the force of steam, raised to the pressure of several atmospheres, but instead of condensing to let the steam escape. I of course answered at once that the loss of power would be one atmosphere, diminished power by the saving of an air-pump with its friction, and in many cases with the raising of condensing water. I never saw a man more delighted, and I believe that within a month several puffers were in actual work.

Davies Gilbert. *"J. S. Enys, Esq."*

Trevethick's steam engine (c. 1802)

Thus, Trevithick started to develop variations on his steam engine design. It was called the "Cornish boiler," a horizontal high-pressure engine incorporating a series of radical improvements. He was using a cylindrical high-pressure wrought iron boiler, in which the furnace and the boiler were combined. This permitted the generation of

Figure 35: Trevithick's Nr. 14 engine (1804).

Source: Wikimedia Commons, Scientific American, Supplement No. 470, 3 January 1885

steam of much higher pressure (around ten times the average working pressure of a contemporary Watt engine). Trevithick also used a strong cast-iron boiler, which was utilized as a structural member. He then placed the cylinder inside it, the great advantage of this being that it kept the cylinder hot and so did not have to waste steam reheating the cylinder with every power stroke. Trevithick was the first to safely take advantage of steam to move a piston at well above atmospheric pressure. This design as such was never patented though.

So, Trevithick's system had a basic design that increased its efficiency because all the heat was radiated directly to the water, and was quite compact because the fireplace was inside the boiler (Figure 35). This all made a more compact construction possible, so "en passant" he discovered mobility for his engine.

> *Trevithick, after two years spent in numerous working experiments, under very trying circumstances—from the want of sufficient money, from the greatly depressed state of the mining interests in Cornwall, and from the disputes and lawsuits which had led mine adventurers and mine engineers to mistrust one another—had satisfied himself that a steam-engine would work without an air-pump or condensing water; that neither beam nor parallel motion, nor foundations of masonry, were absolutely necessary; and that the boiler, for conveniently supplying high-pressure steam, need not be one quarter of the weight, or cost, of the low-pressure boilers then in use, for producing an equal amount of power. He had conveyed an engine from mine to mine in a common cart, at a cost of 10s., and even this expense might have been saved by placing the engine on wheels, and driving them around by the force of the steam* (Trevithick, 1872, p. 105).

Trevithick's steam road carriage

For the application of his engine, on 26 March 1802, Trevithick was granted a patent in cooperation with his cousin Andrew Vivian (1759–1842), and his brother-in-law William West. It was British patent No. 2599: "Steam Engines: Improvements in the Construction thereof

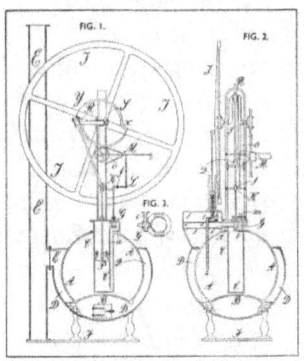

Fig. 89. Trevithick and Vivian's high-pressure steam engine. From their patent specification, 1802

Figure 36: Trevithick's and Vivian's patent (1802).
Source: (Trevithick, 1872, p. 129), also on Grace's Guide: www.gracesguide.co.uk/Richard _Trevithick:_1802_Patent

and Application thereof for Driving Carriages" (Figure 36). Soon he started applying the small steam engine in mobile applications like the stage coach "The London Steam Carriage" (Figure 38).

> While Trevithick and Vivian were in London securing their patent, it was recommended to them that a steam-powered road carriage for the city would be profitable. London then had a population of just over a million, with its industrial heartland surrounded by farms and market gardens supplying fresh food. The limiting factor on expansion was the length of time it took to bring perishable foods like milk from outlying areas to the center. The more quickly foodstuffs could be transported over greater distances, the more the city could grow. Faster travel would also mean better lines of communication for business between London and other towns and cities. Work began on another road carriage, with the wrought iron boiler, cylinder and cast iron parts made at Harvey & Co, supervised by William West, another of Trevithick's brothers-in-law...

> The engine was shipped to London from Falmouth, arriving at William Felton's coach building works at 36 Leather Lane, Holborn in London, in April 1803. Felton made a new carriage capable of carrying up to eight people, which was two more than a standard stagecoach. It had a steerable front wheel moved with a tiller by the driver who sat outside the passenger pod, and two large driving wheels at the rear...West stayed in London for five months to oversee the assembly of carriage and engine, with Trevithick and Vivian visiting at intervals while lodging at 1 Southampton Street in the Strand. The total cost of the steam carriage and engine was about £207.[52] At a trial very early one summer morning, when the streets were clear

Figure 37: Trevithick's common road passenger locomotive: the London Steam Carriage (1803).
Source: Cornish Studies Library © Cornwall County Council, Wikimedia Commons

---

[52] Equivalent in 2010 to £186,000 using average earnings.

of horse-drawn vehicles and pedestrians, the carriage travelled along Tottenham Court Road and City Road, through Oxford Street and back to the coach builder. On another occasion, Vivian steered the carriage from Leather Lane along Gray's Inn Lane to Lord's Cricket Ground, on to Paddington and back again by way of Islington—a round trip of some 16km. It ran at a speed of 13–14km/hour on the flat. The London road trials showed up defects in the firebox design. The motion of the carriage tended to shake the fire bars loose and burning coals dropped into the ash pan. Though the carriage was reportedly seen by "tens of thousands" of spectators during the London trials, the partnership of Trevithick, Vivian and West received no orders for the new vehicle. With money running out and more interest being shown in Trevithick's other engines, the steam carriage was abandoned. The coach body was sold and the engine transferred to a hoop iron rolling mill, where it worked for many years as a stationary engine.[53]

Trevithick's steam railway locomotive

The "Cornish boiler" was originally used as a stationary engine, but it had one huge advantage. It could, due to its compactness, also be used as a mobile engine. So it became used in applications that were mobile, like steam coaches at first and steam locomotives later. Having the idea, he built a prototype. This had to be tested, as happened in the village of Camborne with the first Camborne common road locomotive to be driven by the force of high-pressure steam: the "Captain Dick's Puffer."

And on Christmas-eve, 1801 [they] conveyed the first load of passengers ever moved by the force of steam. The start was from Tyack's smiths' shop, where the smaller parts had been made. East and west ran the great main coach-road to London, on which the Cornish coach, at that time a van or covered wagon, conveyed the few who travelled on wheels. Northwards, towards the great house of Lord Dedunstanville, at Tehidy, the road was more hilly. The south road was a rude country lane, in the worst possible order, with a sharp curve at the commencement, and steeper gradients than either of the other roads…This southern road from Camborne was the worst of the four that were open to Trevithick's choice for testing his first locomotive, carrying as many

---

[53] Text from website Engineering Timelines, Trevithick, First road carriages (accessed June 2014). Source: http://www.engineering-timelines.com/who/Trevithick_R/TrevithickRichard5.asp.

passengers as could find standing-room on it—perhaps half a dozen or half a score. A piece of newly-made road with loose stones, just where the incline increased, and when the small boiler had expended its hoarded stock of high-pressure steam, heaped an insurmountable barrier against the small wheels of the engine, and baffled the engineer for the moment. While the road was being smoothed, the steam had increased its elastic force. Another progress was made, and the first half-mile had been travelled on a steam-horse (Trevithick, 1872, pp. 98-100).

In 1802 their patent was granted. Richard Trevithick (40 percent), Andrew Vivian (40 percent), and William West (20 percent) were partners in the patent. The steam carriage resembled a stagecoach and was upon four wheels. The steam engine had one horizontal cylinder, which, together with the boiler and furnace box, was placed in the rear of the hind axle (Figure 36).

Richard Trevithick tried several times to interest investors in the steam railway locomotive. One of the later attempts was in London in 1808, when he set up a circular demonstration track, round which ran his locomotive "Catch-me-who-can" (Figure 38).

Figure 38: Trevithick's locomotive "Catch me who can" (1808).
Source: Wikimedia Commons

In the year 1808, Trevithick built a railroad in London, on what was known later as Torrington Square, or Suston Square, and set at work a steam carriage, which he called "Catch-me-who-can." This was a very plain and simple machine...This engine, weighing about 10 tons, made from 12 to 15 miles an hour on the circular railway in London, and was said by its builder to be capable of making 20 miles an hour. The engine was finally thrown from the track, after some weeks of work, by the breaking of a rail, and, Trevithick's funds having been expended, it was never replaced...Trevithick applied his high-pressure non-conducting engine not only to locomotives, but to every purpose that opportunity offered him. He put one into the Tredegar Iron Works, to drive the puddle-train, in 1801...In 1803, Trevithick applied his engine to driving rock drills, and three years later made a large contract with the Trinity Board for dredging in the Thames, and constructed steam dredging–machines for the work, of the form which is still most generally used in Great Britain, although rarely seen in the United States the chain-and-bucket dredger...A little later, Trevithick was engaged upon the first and unsuccessful attempt to carry a tunnel under the Thames, at London; but no sooner had that costly scheme been given up, than he returned to his favorite pursuits, and continued his work on interrupted schemes for ship propulsion...Trevithick at last left England, spent some years in South America, and finally returned home and died in extreme poverty, April, 1833, at the age of sixty-two, without having succeeded in accomplishing the general introduction of any of his inventions (Thurston, 1878, pp. 176-177).

As illustrated, Trevithick designed several railroad steam locomotives, such as the Coalbrookdale locomotive for the ironworks at Coalbrookdale in 1804 (Figure 39) and the Trevithick Gateshead/Wylam Colliery locomotive in 1805. They all had problems and broke the rails they were driving on. He was more successful with the Penydarren locomotive for Samuel Homfray, the owner of the Penydarren ironworks near Dowlais, South Wales (1804).

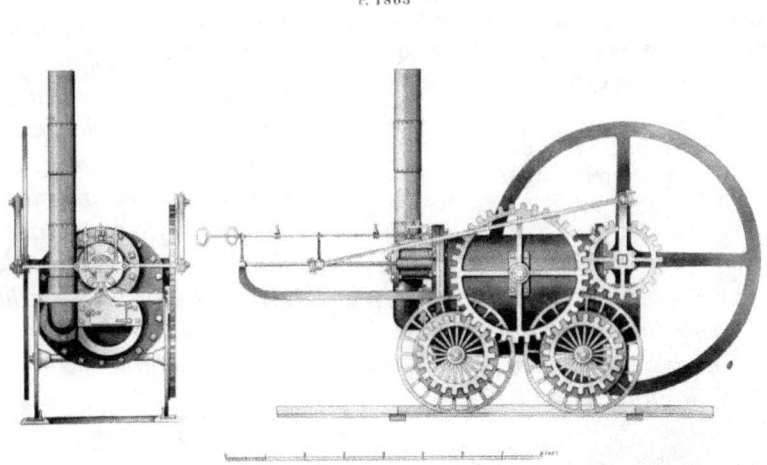

Figure 39: Trevithick's Coalbrookdale locomotive (c.1803).
Source: Wikimedia Commons, The British Railway Locomotive, H.M.S.O., Science Museum

Samuel Homfray, owner of the Penydarren Iron Works, Merthyr Tydfil, made a bet of 1,000 guineas with Richard Crawshay, owner of the Cyfarthfa Iron Works, that he would construct a steam engine to haul a load of 10 tons of iron from his works along the tramway, to Navigation House, Abercynon. The bet was accepted, and the work people became tremendously interested in the event. Homfray had the assistance of Cornishman Richard Trevithick, whose plan for a "High Pressure Tram-Engine" had earned the ironmasters support. Early in 1804 Trevithick's engine, with its single horizontal cylinder, 8 foot flywheel and long piston-rod, was ready, and February the 14th was chosen for the great test. People came from far and near to witness the great experiment.

Figure 40: Trevithick's Penydarren locomotive on its epic journey from Merthyr to Abercynon, 21 February 1804.
Painting by Terence Tenison.
Source: www.museumwales.ac.uk/

The five trams were loaded with the iron, and 70 men added themselves to the load. With shouts of encouragement, the engine started on its journey. Unfortunately disaster soon struck, for the chimney of the locomotive struck a low bridge and both were destroyed. According to the terms of the wager, Trevithick had to control and repair the engine unaided. In a short time he had cleared the debris and repaired the chimney, and soon was careering along at a speed of five miles an hour to his destination at Abercynon, which was reached without further mishap. Due to the Steep gradients and sharp curves the engine failed to make the return journey even though it had no load, but it had been proven that Steam Locomotion was a possibility (George, 2013).

Trevithick's other activities

Trevithick went in 1807 into a partnership with Robert Dickinson. They acquired several patents, but the partnership was not too successful and ended in a bankruptcy (Henry Winram Dickinson & Titley, 1934).

Trevithick in 1807 entered into partnership with Robert Dickinson, a patent speculator and West Indies merchant living in Great Queen Street near Covent Garden in London. This led to several joint patents for some of Trevithick's inventions. He set up a workshop in the yard of the house at 72 Fore Street, Limehouse, to which he moved with his family in 1808. The first of the patents (No. 3148) was taken out on 5 July 1808. It related to Machinery for towing, driving or forcing and discharging ships and other vessels of their cargo...*The idea was to tow ships up rivers, lie alongside wharfs and discharge cargo using a steam windlass, but it was abandoned after protests by ship workers' unions such as the Society of Coal-Whippers. This was followed by patent No. 3172 (31 October 1808) for iron tanks or* Stowing ships' cargoes by means of packages to lessen expense of stowage, and keep the goods safe...*An example was given of transporting whale oil. Ships catch whales and boil the blubber for oil. However, the timber storage casks leak and the wood soaks up a proportion of the oil—and the crew has to wait for the oil to cool before handling it. Iron tanks could be filled with hot oil and wouldn't leak or absorb oil. Metal tanks would also be better for any perishable cargo, such as food or drink. Using iron saved on expensive timber, as suitable cask material was usually imported. The next year, Trevithick and Dickinson took out a single patent (No. 3231) on 29 April 1809 for nine separate inventions for* Naval architecture and navigation applicable to other purposes. *It included iron floating docks, iron ships for ocean service, iron masts and spars, bending timber with steam, diagonal framing for ships, iron buoys, steam engines for general ships use, rowing trunk and steam cooking...Trevithick then turned his attention to the propulsion of sea vessels, with another joint patent on 23 March 1810. This covered* Inventions or new applications of known Powers to propel Ships & other Vessels employed in Navigating the Seas or Inland Navigation to aid the recovery of Shipwrecks, promote the health & comfort of the Mariners and other useful purposes. *The ideas contained in these patents were not all realized. Possibly they could not be, given the technologies available at the time. It's likely that Dickinson provided most of the considerable financial outlay required but this aspect must have preyed on Trevithick's mind too, as he tried to raise money by selling or mortgaging his mine shares and property.*[54]

---

[54] Text from website Engineering Timelines, Trevithick, Dredging, tunneling and more

In Table 2 an overview of Trevithick's patents is shown. It shows both the 1802 to 1816 period of patent activity and the 1827 to 1832 period of patent activity.

Table 2: Overview of Trevithick's patents

Date	Number	Patentee	Description
24 March, 1802	2599	R.Trevithick, Andrew Vivian	Construction of steam engine. Applications to drive carriages and other purposes
5 July, 1808	3148	R.Trevithick, R.Dickinson	Machinery for towing, driving or forcing and discharging ships and other vessels of their cargo
31 October, 1808	3172	R.Trevithick, R.Dickinson	Stowing ship's cargo by means of packages and keep the goods safe
29 April, 1809	3231	R.Trevithick, R.Dickinson	Naval architecture and navigation applicable to other purposes
23 March, 1810	*	R.Trevithick, R.Dickinson	Inventions or new applications of known powers to propel ships & other vessels…
20 November, 1815	3922	R.Trevithick	High pressure steam engine
22 June, 1816	*	R.Trevithick	New apparatus for evaporating water from solutions of vegetable substances
10 November, 1827	*	R.Trevithick	New methods for centering ordnance on pivots,…
27 September, 1828	*	R.Trevithick	Certain new methods of machinery for discharging ships' cargoes & other purposes
27 March, 1829	*	R.Trevithick	A new or improved steam engine
21 February, 1831	6082/ 6083	R.Trevithick	Steam engine (boiler and condenser)
22 September, 1832	6308	R.Trevithick	Steam engine (super heater)

Source: (Henry Winram Dickinson & Titley, 1934, pp. 278-279) * Unknown

> *In May 1810, Trevithick fell ill with typhus, complicated by gastric troubles and "brain fever." Typhus was potentially fatal but despite conflicting advice from several doctors, his natural resilience led to a partial recovery in August. Trevithick returned home to Cornwall in early September 1810 with his eldest son Richard, then almost 12 years old. They travelled by sea and Trevithick had to be carried on board as he was still weak. The*

---

inventions (accessed June 2014). Source: http://www.engineering-timelines.com/who/Trevithick_R/ trevithickRichard7.asp.

ship, a Falmouth Packet, was escorted by a gun brig (Britain was at war with France), evaded a French warship at Dover and reached Falmouth six days after leaving London. Trevithick and his son returned to Penponds near Camborne—apparently walking the entire 26km from Falmouth. He was reunited with his family and learned the sad news that his mother had died in July, when he had been too ill to be told.

By November he had regained full health and was working on his plunger pole engine and other high pressure steam engines. Further misfortune fell the next year. The London Gazette of 5 February 1811 records that Trevithick and Dickinson "being declared Bankrupts are hereby required to surrender themselves to the Commissioners" on the 16th and 23rd February and the 23rd March. The last appointment was delayed until 4th May. The partnership had debts of £4,000 and Trevithick's possessions were seized, leaving him with no option but to take lodgings in a London "sponging house"—somewhere between freedom and a debtors' prison.

After a long struggle, Trevithick was discharged from bankruptcy on 1 January 1814. He had managed to pay back most of the debts—at 16s (80p) in every £1—while his former partner Dickinson paid nothing.[55]

In the later part of his life, Trevithick went to South America; he had sold some of his steam engines to be shipped to Peru for use in the mines of the Pasco Mining Company. It was not without some adventures as the political climate was quite volatile. It was quite an adventurous period, from which he returned in 1827.

In 1816, Trevithick sailed to Peru to sort out problems with some engines he had sold to the silver mines at Cerro de Pasco. When he fell out with the owners, he travelled the country advising other mines and was rewarded when the Peruvian government ceded him some mining rights. He had just begun to operate a copper and silver mine when he was called up to serve in Simon Bolivar's army. He was no soldier, but he did design and build a gun for the rebels before being released back to civilian life. When he returned to his mine, the Spanish army overran the area and he had to flee. Eventually, after 10 years in Peru, Trevithick left for

---

[55] Text from website Engineering Timelines, Trevithick, Dredging, tunneling and more inventions (accessed June 2014).Source: http://www.engineering-timelines.com/who/Trevithick_R/ trevithickRichard7.asp.

> *home and experienced many adventures crossing the jungles of the Isthmus before reaching Colombia, short of funds and half dead. Who should he meet there but Robert Stephenson—the last time they had met was when Trevithick had dandled him on his knee during his talks with Robert's father. Stephenson gave him £50 to pay his passage home. On arrival, Trevithick attempted to resume his engineering career, and the ideas began to flow again—but there was something inevitable about their lack of financial success. A petition to Parliament for a grant for his work in Cornish mines failed, and he died in April, 1833, in Dartford, where he was working on an engine.*[56]

It was clear that Trevithick's stationary version of the high-pressure steam engine had a huge influence on the mine engines, solving their old water-, air-, and transportation problems. Also another important legacy lay in the potential for small, powerful, self-contained engines, particularly in the field of self-propelled transport: the auto locomotive applications. His work laid the foundation for the development of the steam carriage, the steam locomotive, the steamship, the portable engine, the traction engine, and the steam car and lorry, for many of which he built prototypes. He envisioned an unlimited wealth of applications for his machine that not everybody in his time could understand.

> *I have been branded with folly and madness for attempting what the world calls impossibilities, and even from the great engineer, the late Mr. James Watt, who said to an eminent scientific character still living, that I deserved hanging for bringing into use the high-pressure engine. This so far has been my reward from the public; but should this be all, I shall be satisfied by the great secret pleasure and laudable pride that I feel in my own breast from having been the instrument of bringing forward and maturing new principles and new arrangements of boundless value to my country. However much I may be straitened in pecunary circumstances, the great honour of being a useful subject can never be taken from me, which to me far exceeds riches.*[57]

---

[56] Text from the website Cotton Times by Doug Peacock (accessed June 2014). Source: http://www.cottontimes.co.uk/Trevithicko.htm.

[57] In a note to his friend and fellow engineer Davies Gilbert, Trevithick wrote this just before he died. Graces Guide: Life of Richard Trevithick by F. Trevithick: Volume 2: Chapter 27. Source: http://www.gracesguide.co.uk/Life_of_Richard_Trevithick_by_F._Trevithick:_Volume_2:_Chapter_27.

A cluster of innovations

In the preceding we have seen the cluster of innovations around Trevithick's steam engine. The strong patent position of Watt and his commitment to enforce it as he had shown, in combination with the experience the young Trevithick had acquired working for steam engine manufacturers, put him in a unique position. It resulted in a revolutionary design that not only did not need a condenser; its construction was quite compact compared to the steam engines that were being built at that time (both Watt's and Newcomen's design). This resulted in a new field of application, mobile applications, for the steam engine. Although the idea to use a steam engine in a carriage had already been explored by Watt's assistant Murdoch (Figure 46), it was Trevithick's compact design that made the steam locomotives possible. However, being the first to prove that an application is technically feasible does not mean that those designs also are going to be commercially successful. That would take some decades more and other engineers to realize (Figure 41).

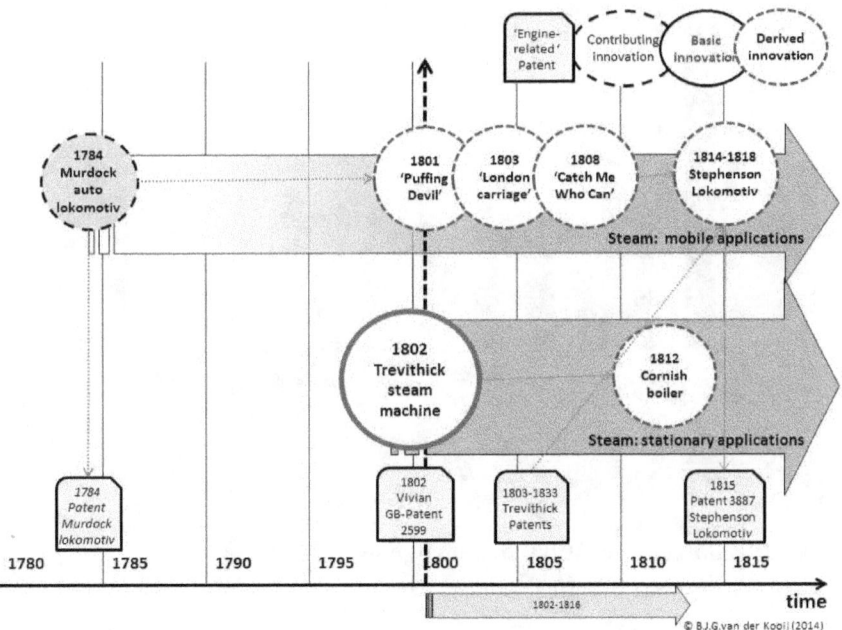

Figure 41: Cluster of innovations around Trevithick's engine.
Source: Figure created by author

Contemporary developments

Trevithick's contributions to the development of the compact, high-pressure steam engine opened a new range of applications. In addition to the stationary applications, now the mobile applications became the focus of attention for many engineers of that time.

Steam engines built in the eighteenth century

As indicated before, the application of the different types of machines went in parallel. In certain areas, because coal was cheaply available, the Newcomen machine was quite popular despite its inefficiency in energy consumption. For the same reason, Watt's machine could be popular at the same time in other regions (Figure 18, Figure 26). Figure 42 shows the manufacturers of the steam engines in use. It clearly shows the dominance of the Newcomen's steam engine in the application of pumping water, as Watt's machine dominated in rotary applications. The steam engines made by Newcomen and Watt & Boulton were quite large and used in static applications.

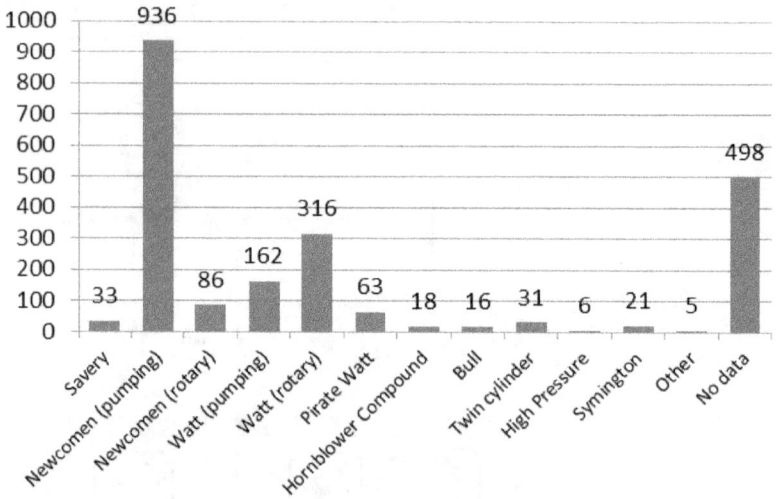

Figure 42: Eighteenth-century steam engines by manufacturer.
Source: (Kanefsky & Robey, 1980, p. 182), Type: Table 3

As these static steam engines (that is, not used in rotary applications like the locomotive) were used from the early 1700s up to the 1800s, one could wonder how many of these machines were made and what they were used for. Figure 43 shows the number of steam engines by

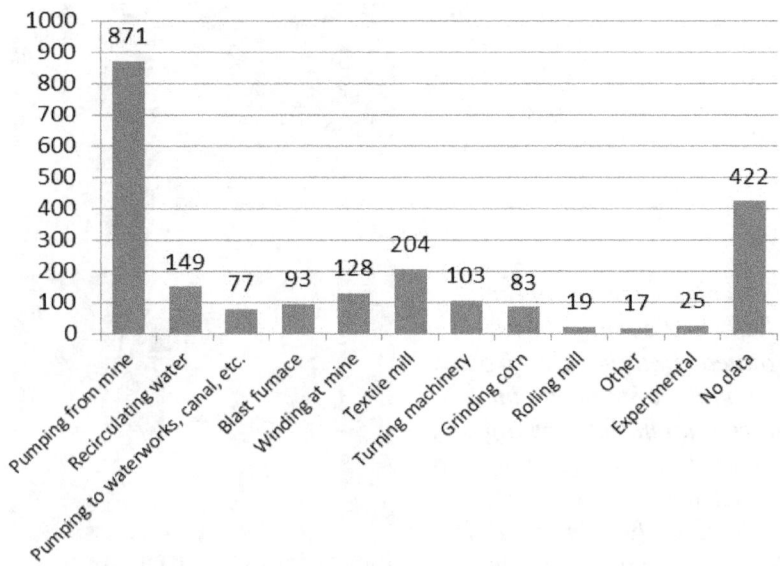

Figure 43: Eighteenth-century steam engines by application.
Source: (Kanefsky & Robey, 1980, p. 182) Application: Table 7.

application in the eighteenth century. Their application in pumping water out of mines is quite dominant.

Stationary applications for steam engines

As can be seen in Figure 44 and Figure 45, the steam engine was the source of power for a range of turning machines, that is, machines and tools that needed rotary energy to fulfill their tasks (like a spinning machine). As it was impossible and uneconomical to apply small steam engines for each machine, the central steam engine's rotative energy had to be distributed. So, for the distribution of rotary power from the source (the steam engine) to the target (the machine or tool), a "power distribution system" of line shafts, pulleys, and belts was created (Figure 45). These line shafts—usually three inches in diameter—were suspended from the ceiling and extended the entire length of the factory floor. From the centrally located line shafts suspended from the ceiling of the factory, the individual machines could be powered by leather belts.

This system was widely applied in woodworking shops, machine shops, sawmills, gristmills, and textile mills. Obviously in those application fields where quite a lot and some different machines were used, the line shaft and belt system was popular. Not only in noisy and dirty factories (Figure 45) but also in workshops, machines were used for more delicate tasks such as jewelry manufacturing (Figure 44). However, the system had some drawbacks.

Figure 45: Belt-driven power distribution in manufacturing (machine shop in Knight Foundry, Sutter Creek, Amador County, California).

Source: Library of Congress Prints and Photographs Division, Washington, DC 20540, http://hdl.loc.gov/loc.pnp/ pp.print

The entire network of line shafts and countershafts rotated continuously—from the time the steam engine was started up in the morning until it was shut down at night—no matter how many machines were actually being used. If a line shaft or the steam engine broke down, production ceased in a whole room of machines or even in the entire factory until repairs were made...maintenance tasks took significant amounts of time, as a large plant often contained thousands of feet of shafting and belts and thousands of drip oilers (Devine, 1983, p. 352).

Figure 44: Belt-driven power distribution for individual tools.

Source: Wikimedia Commons, Jewellery Quarter Museum, www.bmag.org.uk

Mobile applications for steam engines

The compact Trevithick high-pressure steam engine was growing more mature in stationary applications (Figure 35). Soon it also became available as a new source of rotary power. The combination of the compact high-pressure boiler, the piston, and the crank supplied the rotary motion that was needed in a lot of mobile transport applications, for example, in the transportation of products and materials that was done with horse-pulled carriages or horse-driven passenger carriages, the traditional coaches that had been the only means of transportation for centuries. All these applications were faced with the limitations of their source of energy, the horse. As beautiful an animal the noble horse can be, it was maintenance-intensive (requiring a stable and hay), supplied a limited amount of energy (one horse power each), and polluted the streets with its manure. A horse did not last long, as it had to be replaced regularly on longer voyages.

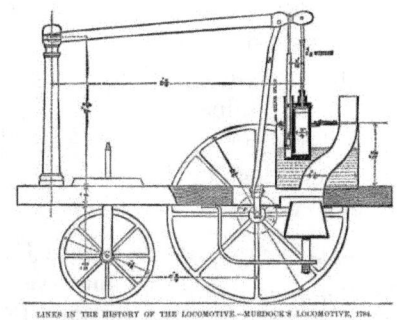

Figure 46: Murdock's design for an auto locomotive (1784).
Source: Wikimedia Commons

Steam carriage

It is not too difficult to understand that the application of the steam engine in mobile applications was envisioned by many other inventors involved in the steam machine. Take, for example, John Murdock, working for Watt and Boulton, who designed a steam carriage as early as 1784 and patented it (Figure 46). James Watt himself, then occupied in perfecting his steam engines and in a different phase of his life, did oppose the further development of his idea. Trevithick, soon after the development of the stationary compact high-pressure steam engine, patented his idea for a steam engine-powered passenger carriage (see Figure 37) in 1803, as soon as his high-pressure steam engine locomotive proved to be feasible. But there were more applications where steam power could replace horse power, for example, in agriculture as the "steam tractor."

Steam tractor

For centuries manpower and animal power had been the primary sources of energy in agriculture. With the arrival of steam power in the more compact version of Trevithick's machines, this changed. The steam tractor became the workhorse for agricultural applications such as ploughing, grinding, and transporting harvest (Figure 47). Not only did the use of the steam tractor have a dramatic effect on mechanical farming, but the manufacturing of the steam tractor itself became an important activity. In England John Fowler developed a system in the 1850s that featured a stationary steam engine that pulled plows by means of wire ropes: the "Roundabout system."

Figure 47: Line drawing of a Burrell universal-type ploughing engine from around 1890.
Source: Wikimedia Commons

Between 1830 and 1850 patents describing steam cultivating machinery were numerous. Probably the most noteworthy innovators being John Heathcoat of Tiverton, Lord Willoughby de Eresby, and the Marquis of Tweeddale. In 1854 the Royal Agricultural Society of England (RASE) offered a prize of £500 for "the steam cultivator which shall in the most efficient manner turn over the soil and be an economical substitute for the plough or spade." The first contest took place at the Society's summer meeting in Carlisle in 1855, and was repeated at Chelmsford in 1856 and Salisbury in 1857, without an award being made. The judges however awarded a medal to John Fowler as a reward for his strenuous endeavours, adding, "Steam ploughing as such had attained a degree of excellence comparable in point of execution with the best horse work."...After 1865 most steam ploughing and cultivation was undertaken by steam ploughing contractors. Many farmers, other than the largest landowners, found the initial cost too high to justify investment. The travelling contractors were a unique breed. Each set of tackle usually comprised four men and a boy living together in a living van which travelled with the engines, implements and water cart.[58]

The mobile steam tractor as we see it in use today—moving across the fields pulling a plough—took quite a time to develop, as the heavy machines had problems working on the soft soils. In the United States, the self-propelled and self-steering tractor occurred in the late 1870s.

---

[58] Text from website Steam Plough Club; History of steam ploughing (accessed June 2014). Source: www.steamploughclub.org.uk/history.htm.

Steam boats

Near the end of the eighteenth century, the transportation of coal from the mines, being quite voluminous, was a labor-intensive task. Canals were built for the transportation barges pulled by horses. The first canal, developed by the Duke of Bridgewater between the coal mines in Worsley, Lancashire, and the big market of Manchester—the Bridgewater Canal—was nearly six miles long (Figure 48). It created a small revolution among other mine owners when the dropping prices of coal created a steadily increasing demand for coals to heat and cook in the cities, making the 'Canal Duke' quite wealthy[59].

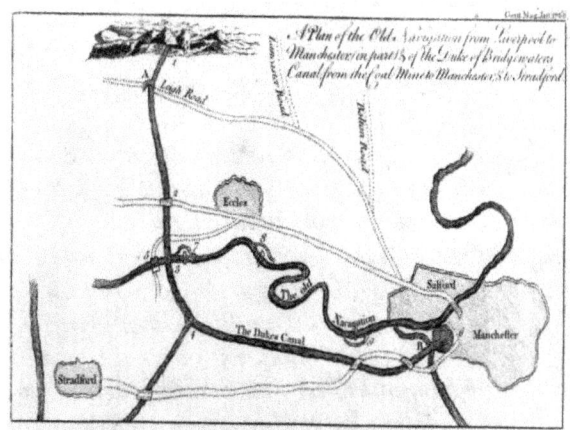

Figure 48: Plan for the Bridgewater Canal (1776).

Source: Gentleman's Magazine, www.virtualwaterways.co.uk/featuresuploads/11b.jpg

> Soon canals were constructed everywhere in England. The earlier opposition and distrust that were encountered when the Duke of Bridgewater's Canal was created in 1767, the Grand Trunk Canal in 1772, and other later canals, had given place, owing to the financial results achieved, to the wildest speculation. Numberless worthless schemes were launched upon. In the course of the four years ending in 1794, not fewer than eighty one Canal and Navigation Acts were obtained; of these, forty-five were passed in the two latter years, authorizing the expenditure of not less than £5.3 million[60] (H.W. Dickinson, 1913, p. 58)

It was quite logical that inventors tried to apply the mobile steam machine in boats that could pull barges. An early example can be found

---

[59] For more detail on the wealth of the Canal Duke, see: Falk, B.: The Bridgewater Millions: A Candid Family Story. Hutchinson & Company Limited (1942). And: Malet, H.: Bridgewater, the Canal Duke, 1736-1803. Manchester University Press (1977).
[60] Equivalent in 2010 to £6.19 billion using average earnings.

in the steamboat that William Symington constructed for Lord Dundas and Lord Bridgewater (Figure 31).

> *Symington had the misfortune of losing the co-operation of Mr Miller, who, most unaccountably, at once and for ever abandoned experiments in steam navigation. From that time, until the year 1800, this invaluable nautical auxiliary was allowed to lie dormant, the state of its inventor's pecuniary resources being such as to prevent his attempting to carry it further unaided. One day, however, while going to examine a field of coal he intended to rent or purchase, he heard someone calling to him, and, on looking round, saw Lord Dundas beckoning to him from the window of his carriage, which had just passed. On going to the carriage, his lordship told him that, having seen his former steamboat experiment, he had come down from London principally for the purpose of seeing him, in order to learn whether steamboats could not be substituted for the horses used in dragging vessels on the Forth and Clyde canal, of which his lordship was a large proprietor and governor. Mr. Symington, fortunately for his country and the world, although most unfortunately for himself and family, gave up all thoughts of the colliery, and returned home, elated with the thought of being able to re-embark in his favorite project under such promising auspices. On subsequently waiting on his lordship by appointment, an arrangement was speedily effected, and, in 1801, the first boat, named the "Charlotte Dundas" (in honor of his lordship's daughter, afterwards Lady Milton), was built for the express purpose of being propelled by a steam engine. After making a trip to Glasgow, she was set to work, and towed on various occasions vessels in the canal, besides running down into the river Forth and dragging thence at one time up the river Carron into the canal at Grangemouth, four or five sloops, detained by a contrary wind. Although thus far successful, the proprietors of the canal, with the exception of Lord Dundas, fearing its banks might be injured by the undulations caused by the paddle-wheels, ordered it to be discontinued. His Lordship, however, who was not so easily prejudiced or discouraged, advised Mr. Symington to get a model of his boat constructed, and take it to London—an advice which was followed by Mr. Symington himself taking the model to Arlington Street (No. 17), and presenting it to his Lordship, who was so much pleased with it that he introduced him to his Grace the Duke of Bridgewater, who not only expressed his admiration of the plan, but immediately gave orders that eight boats of similar*

construction should be got ready as speedily as possible for his canal. Soon after his interview with the Duke of Bridgewater, Mr. Symington returned to Scotland, and completed his second and largest steamboat, likewise named the Charlotte Dundas, for Lord Dundas. This boat was tried in March 1803, when she towed two laden sloops, the Active and Euphemia, of seventy tons burthen each, from Lock No. 20 to Port Dundas, Glasgow, 19 miles in six hours, notwithstanding that during the whole time so strong an adverse gale prevailed that no other vessel in the canal could that day move to windward.[61]

The steam engine-powered boats soon became an alternative for another old source of energy: wind power. The Comet, which was designed and commissioned by Henry Bell, combined sails and a steam engine and sailed its inaugural voyage in 1812. After that it maintained a regular "steam-powered" passenger service between Glasgow, Greenock, and Helensburgh. No longer did ferries need to be so dependent upon wind and tide.

Steam locomotive

In addition to the earlier mentioned efforts of William Trevithick to create a steam locomotive, many other steam locomotives were designed as the advantages of the steam power-driven machines became obvious. A well-known example of one of these steam locomotives was the "Rocket," developed by father George Stephenson (1781–1848) and son Robert Stephenson (1803–1859) in 1829; the Rocket took part in the Rainhill Trials and eventually became the winner. These Rainhill Trials were held to select a steam locomotive for the Stockton & Darlington Railway, a project that was not without obstacles and problems of its own, even before construction, as it

Figure 49: The steam powered sailboat 'Comet' (1812).
Source: Gentleman's Magazine, www.virtualwaterways.co.uk/featuresuploads/11b.jpg

---

[61] R. Bowie: *A Brief Narrative, Proving the Right of the Late William Symington...to be Considered the Inventor of Steam Land Carriage Locomotion.* Sherwood, Gilbert and Piper, 1833. Source:
http://www.electricscotland.com/history/men/biographyofwilli00rank.pdf.
See also: B. E. G. Clark: *Symington and the Steamboat.* Lulu.com, 2010. 144 pages.

had to have permission in the form of an Act of Parliament before it could start (Echo, 2008)

> *In those days, before the birth of the railways in 1825, coal was hoofed over the hills of County Durham to the sea. It was carried in panniers slung over the backs of packhorses and trudged along routes such as Carmel (Coal) Road in Darlington. So, to get the coal from the inland mines to coast in less cumbersome and costly fashion, several plans for canals—it was the time of the Canal Mania—developed. Because Stockton had plans for a new canal project (the Stockton and Auckland Canal), Darlington was in danger of missing the boat. So they proposed a railway project and created a company, the Stockton & Darlington Railway company, and raised enough money to fund it (£125,000 which would be £88,7 million using average earnings in 2010). The Stockton committee failed to raise enough money for their canal and abandoned it for a railway project of their own. The next obstacle was the proposed trajectory that ran over the land of Lord Elton of Windleston Hall and Lord Darlington. Lord Darlington lived purely for his fox-hunting. He was obsessed by it. Yet the planned railway ran over his land. In fact, in their haste to get their plan before Parliament ahead of Stockton, the Darlington contingent had driven the railway through Lord Darlington's fox coverts—specially planted thickets where foxes lived until his lordship and his dogs rooted them out. With Lord Darlington implacably opposed to the railway, the Darlington plan began 1819 staring defeat in the face...They bought off Lord Eldon— actually they bought three-and-a-quarter acres of the Windlestone estate from Lord Eldon at such a generous price that this lordship forgot his opposition—and Overton engineered a line which was nine miles shorter and spared the fox coverts of Lord Darlington...Then, on January 29, 1820, they were stopped dead in their tracks. King George III died. He had reigned for 60 years—not all of them with his marbles in strictly the right places—but inconsiderately chose this year of all years to depart. And with him went the House of Commons. New king, new elections. There would be no Parliamentary action until 1821...So they reorganized themselves, made a new trajectory for the project, and repeated their lobbying. They even went so far as to try to get their own MPs returned to the House in the 1820 General Election...In February 1821, the Bill began its passage through Parliament. First Reading (February 20)—no problem. Second Reading (February 28)—no problem. On to the Committee*

Stage. But then the railway solicitor, Francis Mewburn, read the Parliamentary small print. Before a Bill could be considered in committee, 80 per cent of the money supporting it had to be raised. In his hotel room in London, Mewburn did his mathematics and discovered that the pioneers were £7,000 short [about £5 million in 2010]. But the money was not to be found. Mewburn wrote to Edward Pease [a Quaker who had cofounded the Stockton & Darlington Railway company] back home in Northgate, Darlington, that if the £7,000 was not forthcoming within three days, he would return North and all would be lost. Pease saved the day. He threw in the £7,000 (worth today about £225,000) from his personal savings...The Bill passed the Committee Stage "in high style." It whipped through its Third Reading in the Commons on April 12, stormed through the House of Lords on April 17, and on April 19 [1821] it received its Royal Assent from George IV...On July 23, the committee formally decided that it was to be a railway, not a tram road.[62]

So the Stockton & Darlington Railway project could start (Figure 50). It was planned for the transportation of coal from the local mines by a railway. How would they power the transportation? By horse? By ropes connected to stationary steam engines? But then George Stephenson met with Pease. It was a meeting that would start Stephenson's ascent and earn him a fortune. (When he died twenty-seven years later, he left £140,000.[63])

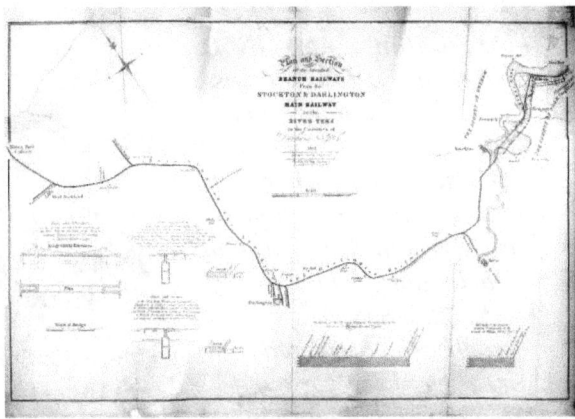

Figure 50: Stockton and Darlington railway plan (1827).
Source: Wikimedia Commons

---

[62] Text from: Stockton & Darlington Railway: history of the world's first passenger railway. *The Northern Echo*, June 2008. Source: http://www.thenorthernecho.co.uk/history/ railway/stockton/.
[63] Ibid. Kitchen table talks changed railway history. Equivalent to about £100 million in 2010.

George Stephenson (1781–1848) was the son of a colliery fireman and grew up living next to the Wylam Wagonway (a five-mile wooden wagonway that had been built in 1748 to take the coal from Wylam to the River Tyne) and grew up with a keen interest in machines. In 1802 Stephenson became a colliery engineman. When he was twenty-seven, Stephenson found employment as an engineman at Killingworth Colliery. Every Saturday he took the engines to pieces in order to understand how they were constructed. This included machines made by Thomas Newcomen and James Watt. By 1812 Stephenson's knowledge of engines resulted in him being employed as the colliery's enginewright...In 1813 Stephenson became aware of attempts by William Hedley and Timothy Hackworth, at Wylam Colliery, to develop a locomotive. Stephenson successfully convinced his colliery manager to allow him to try to produce a steam-powered machine. By 1814 he had constructed a locomotive that could pull thirty tons up a hill at 4 mph. Stephenson called his locomotive, the Blutcher...

On the 19th of April 1821 an Act of Parliament was passed that authorized a company owned by Edward Pearse to build a horse railway that would link the collieries in West Durham, Darlington and the River Tees at Stockton. Stephenson arranged a meeting with Pease and suggested that he should consider building a

Figure 51: Stephenson's "Blutcher" (1814).

Source: Wikimedia Commons

locomotive railway. Stephenson told Pease that "a horse on an iron road would draw ten tons for one ton on a common road." Stephenson added that the Blutcher locomotive that he had built at Killingworth was "worth fifty horses." That summer Edward Pease took up Stephenson's invitation to visit Killingworth Colliery. When Pease saw the Blutcher at work he realised George Stephenson was right and offered him the post as the chief engineer of the Stockton & Darlington Company. It was now necessary for Pease to apply for a further Act of Parliament. This time a clause was added that stated that Parliament gave permission for the company "to make and erect locomotive or moveable engines." In 1823 Edward Pease joined with Michael Longdridge, George Stephenson and his son Robert Stephenson, to form a company to make the locomotives [The Robert Stephenson & Company]. The Stockton & Darlington line was opened on 27 September 1825. Large crowds saw George Stephenson at the controls of the Locomotion as it pulled 36 wagons filled with sacks of coal and flour. The initial journey of just less than 9 miles took two hours.[64]

Rainhill Trials[65]

As the creation of railroads and railway enterprises for the transportation of goods and materials was a hot item in those days, the Liverpool-Manchester Railway was at first intended for the transportation of coal. So, to select a steam locomotive for the Liverpool & Manchester Railway, under construction, in 1829 a competition was held. One of the rules stipulated:

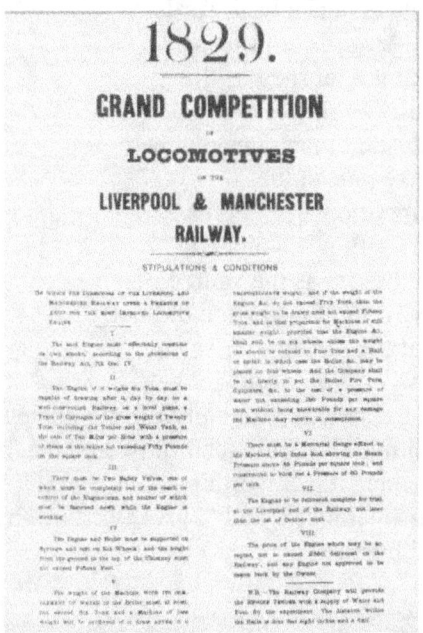

"The Engine, if it weighs Six Tons, must be capable of drawing after it, day by day, on a well-constructed Railway, on a level plane, a Train of Carriages of the gross weight of Twenty Tons, including the Tender and Water Tank, at the rate of Ten Miles per Hour, with a pressure of steam in the boiler not exceeding Fifty Pounds on the square inch."

Figure 52: Stipulation & Conditions for the Grand Competition.
Source: http://apps.robertstephensontrust.com/Blog/?c=rainhill&p=2

---

[64] Text from website Spartacus International; George Stephenson. Source: http://spartacus-educational.com/RAstephensonG.htm. (Retrieved June 2014).
[65] Text based on: Mechanics Magazine. No. 322, 323, 324, 325, October, 1829. Source: http://www.resco.co.uk/rainhill/rain.html; Also The Rainhill Trials. Source: http://www.rainhilltrials.com/userfiles/File/The_Spectacle.pdf (Retrieved June 2014).

The trials were held over a number of weeks. Grandstands were erected, and many sightseers came to watch the events on 6–14 October 1829. Somewhere between 10,000 and 15,000 people were there to see the first day of the trials—quite impressive numbers when you consider their journey there by road in every type of vehicle. Large numbers came from Liverpool, Warrington, St. Helens, Manchester, and the surrounding areas. The sheer numbers that visited indicate the tremendous and widespread interest in the event.

The tracks used for the trial were a little more than a mile and a half long, so each engine had to travel the whole distance backwards and forwards ten times, making a journey of thirty miles. Originally five contestant machines planned to take part in the competition. As Thomas Brandreth's "Cyclopede" used horses as the source of energy, the Cyclopede did not merit the serious consideration of the judges

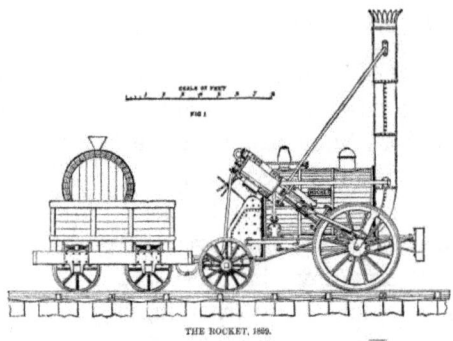

Figure 53: Stephenson's "Rocket" (1829).
Source: Wikimedia Commons, www.stanleys-steamers.gen.nz/images/t11-PLAN.png

and did not—fortunately for the horses—get involved in the competitive running. The four steam-powered machines were Stephenson's "Rocket," Hackworth's "Sans Pareil," Timothy Burstall's "Perseverance," and "Novelty," built by John Ericsson and John Braithwaite.

Figure 54: Ericsson & Braithwaite's "Novelty" (1829).
Source: Wikimedia Commons

Then came the first day of the trial. As the Perseverance was damaged on the way to the trials and Burstall had to spend time trying to repair his locomotive, it only ran on the sixth and final day of the trials. Then it only achieved a speed of 6 mph and was awarded a consolation prize of £25.

Novelty also ran into problems, as it suffered damage to a boiler pipe that could not be fixed properly on-site. Sans Pareil (Figure 56), being too overweight, completed eight trips before cracking a cylinder. Only Stephenson's Rocket completed the 80km round-trip under load, averaging twelve miles per hour while hauling thirteen tons. The prize was £500 (equivalent in 2010 to £349,000 based on average earnings).

Figure 55: Timothy Hackworth's "Sans Pareil" (1829).

Source: Wikimedia Commons, www.railalbum.co.uk/early-railways/sans-pareil.htm

After the trials, the Liverpool & Manchester Railway bought Sans Pareil as well as Rocket. The Liverpool & Manchester Railway was opened on 15 September 1830. It was the first full-scale inter-city railway exclusively powered by locomotives and providing a service for both passengers and freight. Its double track throughout and its strict timetable formed the prototype for subsequent railways throughout the world.[66]

Steam-powered coaches

As steam locomotives were designed on tracks limiting their flexibility of going somewhere, soon the application of steam power to road-based carriages started. These were free to roam everywhere there was a suitable road and the

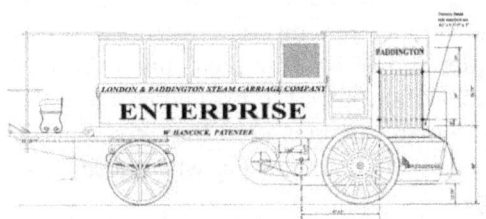

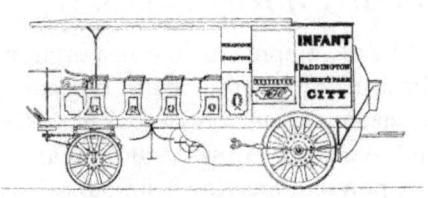

Figure 56: Walter Hancock's steam coach "Infant" (1831) and "Enterprise" (1833).

Source: www.steamcar.net (Enterprise), www.gracesguide.co.uk (Infant)

---

[66] Text source: Wikipedia, Rainhill Trials. Source: http://www.rainhill-civic-society.org.uk/html/ newrainhillHistory.html. (Retrieved June 2014).

horses could provide enough traction power. So, steam-powered stagecoaches soon were developed, for example, by Timothy Burstall, who took part in the Rainhill Trials with the "Perseverance" (it could be a normal coach with the horse missing). Another developer was Walter Hancock (1799–1852), who was also one of the early builders of steam-powered coaches. In 1829 he built a small ten-seater bus called the "Infant," with which in 1831 he began a regular service between Stratford and London. In 1833 this was followed by the "Enterprise," which ran a regular service between London Wall and Paddington via Islington (Figure 55). It was the first regular steam-carriage service and was the first mechanically propelled vehicle specially designed for omnibus work to be operated.

The steam coaches were opposed, however, by the establishment at that time, which had an interest to conserve the existing situation with public horse-driven coaches. Actions ranged from pure sabotage and bribing members of Parliament to support legislation hampering steam vehicle operation, to toll roads charging six to seven times the normal fee (a charge of £2[67] was levied on each steam-carriage journey, whilst the toll for a horse-drawn carriage was 2 shillings). So one could say that these innovations did not face the most favorable circumstances when they were introduced to society. Take, for example, the Locomotive Act of 1861, which imposed restrictive speed limits to 8 km/h in towns and cities and 16 km/h in the country. Or the Red Flag Act of 1865, which reduced the speed limits and required that a vehicle had to be preceded by a man carrying a red flag. It also gave local authorities the power to specify the hours during which a steam-powered vehicle might use the roads.

Mobility infrastructures

The developments in transportation were related to both the infrastructure (roads, canals, railroads) and the transportation systems (the steam-driven vehicles, as described before). They developed hand in hand. As better infrastructure became available, transportation became more popular, creating a stronger demand for transportation vehicles. As technology developed, better (more reliable, more efficient, cheaper) steam engines became available. The developments in the infrastructure followed the previously described developments in the engines and

---

[67] Equivalent to ca. £140 in 2010 based on a real price calculation. Source: www.measuring worth.com.

machines. To give a glimpse, as they deserve another case study, consider the following.

From wagonway to railroad

As stated before, transportation of voluminous and heavy cargo and people was done by boats on the rivers and the newly constructed canals. Also, the not too perfect local roads were used for horse-pulled carriages transporting persons and goods—carriages that were creating deep tracks and potholes, resulting in deteriorating the road condition, especially on the industrial haulage routes.

Something had to be done, and this resulted in strengthening the road surface with wood. These so-called "wagonways" with their "rails" made of wood improved coal transport by allowing one horse to deliver an approximate fourfold increase over the earlier horse-pulled carriages. Wagonways (Figure 57) were usually designed to carry the fully loaded wagons downhill to a canal or boat dock and then return the empty wagons back to the mine. As early as the seventeenth century, wagonways—such as the Wollaton Wagonway built to transport coal from the mines at Strelley to Wollaton, just west of Nottingham, England—were in existence (Lewis, 2006).

Figure 57: A typical wagonway: the Denby plateway at Coxbench.
Source: http://chasewaterstuff.wordpress.com/2013/05/11/some-early-lines-early-tramroads-and-plateways/

Wood proved not to be too durable, and soon iron rods were used to create a stable support for the horse-pulled carriages. In the late 1760s, the Coalbrookdale Company began to fix plates of cast iron to the upper surface of the wooden rails. The replacement of the horse by a traction engine was the next step; the carriages were now pulled by a steam-driven locomotives that were running on the "railway." For the construction in 1758 of the Middleton Railway, which carried coal cheaply from the Middleton pits to Casson Close in Leeds, powers were granted by an Act of Parliament[68].

---

[68] Source: 31 Geo.2, c.xxii, 9 June 1758: An ACT for Establishing Agreement made between Charles Brandling, Esquire, and other Persons, Proprietors of Lands, for laying

From freight to passengers

At first the locally based iron railways were to realize transport of goods an materials (i.e. coal) over shorter distances, for example, between mine and harbor. Soon they were applied over longer distances. In 1825 the Stockton & Darlington Railway (about 25 miles) opened (Kirby, 2002), followed by the Liverpool & Manchester Railway (30 miles) five years later (Thomas, 1980). They were isolated tracks that just connected the cities mentioned. The Stockton & Darlington Railway demonstrated the feasibility of steam-locomotive propulsion. Trains, originally intended for the transportation of goods, soon took on passengers. It proved to be a profitable business, which expanded rapidly as more and more railways connected cities.

It also resulted in the "rail" systems in big cities such as London, where the horse-pulled passenger tramway became popular in 1860 when a horse tramway began operating along Victoria Street in Westminster. In 1870 the Tramways Act passed the British Parliament, granting private operators a twenty-one-year protection on their projects. It was followed by the Light Railways Act in 1896. This latter resulted in dozens of private initiatives to create a tramway. The horses originally used to pull the carriages were soon replaced by steam locomotives.

Later the local rail companies started creating connections between the till-then isolated lines that developed during the railway boom of the 1840s, thus creating a national network, although it was still run by dozens of competing companies. Now passengers and goods could travel over longer distances. Ultimately, it was these networks of interconnected railways that created the infrastructure for mobility, not only for the transportation of goods, but also the transportation of persons.

Figure 58: Horse-power tram of the London Tramways Company (1890).
Source: Wikimedia Commons

---

down a Wagon-Way in order for the better supplying the Town and Neighborhood of Leeds in the County of York, with Coals.

Conclusion

The discovery and application of the steam engine certainly could be an invention if it had been the act of one person. That was not the case, however, as shown in the preceding overview. It was a range of discoveries that started with the "engine of fire" and at the end created the compact high-pressure steam engine. It took a while to go from Savery's engine (let's say 1700) to Trevithick's engine (ca. 1800), events that resulted in the development of the steam engines themselves as well as the range of stationary and mobile applications for the steam engine: from pumps for draining and ventilating the mines, to steam-powered locomotives, steam-powered ships, even the steam bicycle and the steam automobile. Steam power replaced manpower, wind power, waterpower,

Table 3: The important patents in the development of the steam engine and its applications (1698–1802)

Patent Number	Year	Patentee	Invention
356	1698	Thomas Savery	Steam engine (Newcomen)
913	1769	James Watt	Separate condenser
931	1769	Richard Arkwright	Water Frame
962	1770	James Hargreaves	Spinning Jenny
1063	1774	John Wilkinson	Boring Machine
1351	1783	Henry Cort	Iron Making
1470	1785	Edmund Cartwright	Power loom
1565	1786	Edmund Cartwright	Power loom
1876	1792	Edmund Cartwright	Power loom
2045	1795	Joseph Bramah	Hydraulic engine
2599	1802	Andrew Vivian	High pressure steam engine (Trevithick)

Source: (Alessandro Nuvolari & Tartari, 2011, p. Table 2; adapted)

and animal power. It started the second Industrial Revolution—and the Industrial Revolution was one of the greatest discontinuities in our recent history.

It might have been started by Savery's steam pump, based on contributions by many scientists and (hydraulic) engineers (Figure 59). Certainly the work of Thomas Newcomen created a marking stone as he introduced the "steam engine" with Papin's plunger cylinder. The contributions made by James Watt as the inventor, and Matthew Boulton as the entrepreneur, over a range of decades were quite significant. Improvements were made on his machine by numerous others. Then, again, the discovery of the high-pressure steam engine by Trevithick contributed enormously.

So, in total, it was a range of discoveries, small and big, that created at last the "steam engine" that was going to be used as a source of rotary power in applications such as the steam locomotive and the steamboat. It was not as if there was, at a certain moment, an "invention" created by one person. Maybe that moment on the Green of Glasgow that fine Sabbath afternoon when Watt got the idea for the condenser was

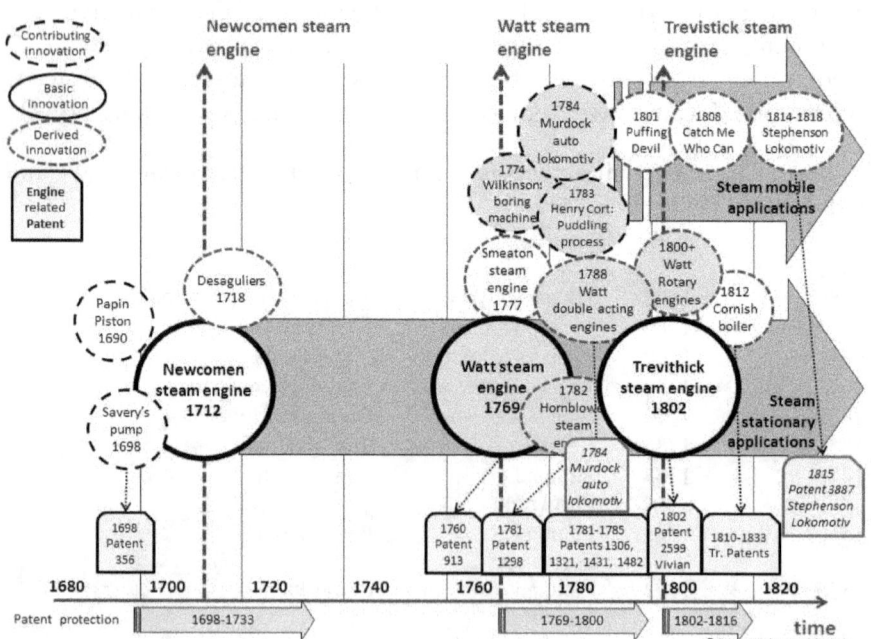

Figure 59: Overview of major innovations in the development of the steam engine.
Source: Figure created by author

important, but such a moment happened for each and every discovery that was made. One can be sure that Savery, Newcomen, Trevithick, and all the others also had similar sparks of insight and creativity. As the discovery of the condenser principle proved important for the many applications that resulted from this discovery, one could say it had a certain fundamental aspect; thus, it can be considered as a basic innovation (like Newcomen's machine and Trevithick's machine). From these basic innovations, the creative individuals involved created a range of derived discoveries, which can be considered incremental innovations when they were applied.

But it is the totality of all those innovations, realized by different inventive people, that certainly can be considered as an "invention." It is thus the totality of three distinct clusters of innovations (Figure 59) that can be called "the invention of the steam engine."

References

Allen, R. C. (2012). Backward into the future: The shift to coal and implications for the next energy transition. *Energy Policy, 50*(November 2012).

Arnold, A., & McCartney, S. (2011). 'Veritable gold mines before the arrival of railway competition': but did dividends signal rates of return in the English canal industry? *The Economic History Review, 64*(1), 214-236.

Bohning, J. J. (1999). The Chemical Revolution. In A. C. Society (Ed.), *An International Historic Chemical Landmark*. Paris, France. : American Chemical Society.

Bourdenet, N. (2003). The Mines Act, 1842. *L'Unité de Formation et de Recherche (UFR) d'Etudes Anglo-Américaines de l'Université Paris X Nanterre*. http://anglais.u-paris10.fr/spip.php?article88

Canny, N. P. (1982). *The Upstart Earl: A Study of the Social and Mental World of Richard Boyle, First Earl of Cork 1566-1643*: Cambridge University Press.

Carnot, S. (1824). *Reflections on the motive power of fire and on machines fitted to develop that power*. Paris: Bachelier, Libraire.

Cheng, K. C. (1992). Historical Development of the Theory of Heat and Thermodynamics: Review and Some Observations. *Heat Transfer Engineering, 13*(3), 19-37. doi: 10.1080/01457639208939779

Conant, J. B. N., L.K. (1948). The overthrow of the Phlogiston Theory *Harvard Case histories in experimental science* (pp. 67-115). Cambridge, Massachusetts: Harvard University Press.

Corfield, B. (2013). Thomas Newcomen the Man. *International Journal for the History of Engineering & Technology, 83*(2), 209-221.

Cutler, D. M., Deaton, A. S., & Lleras-Muney, A. (2006). The determinants of mortality. http://dash.harvard.edu/bitstream/handle/1/2640588/cutler_determinants.pdf

Devine, W. D., Jr. (1983). From Shafts to Wires: Historical Perspective on Electrification. *The Journal of Economic History, 43*(2), 347-372. doi: 10.2307/2120827

Dickinson, H. W. (1913). *Robert Fulton, engineer and artist: his life and works*: John Lane.

Dickinson, H. W. (1939). *A short history of the steam engine*: Cambridge University Press.

Dickinson, H. W. (1947). Tercentenary of Denis Papin. *Nature, 160*(4065), 422.

Dickinson, H. W., & Titley, A. (1934). *Richard Trevithick: the engineer and the man*: Cambridge University Press.

Dickson, P. G. M., & Beckett, J. V. (2001). The Finances of the Dukes of Chandos: Aristocratic Inheritance, Marriage, and Debt in Eighteenth-Century England. *Huntington Library Quarterly, 64*(3/4), 309-355. doi: 10.2307/3817916

Dircks, H., & Worcester, E. S. (1865). *The Life, Times, and Scientific Labours of the Second Marquis of Worcester: To which is Added, a Reprint of His Century of Inventions, 1663, with a Commentary Thereon*: Quaritch.

Echo, T. N. (2008). Effort that kept the mines afloat. *The Northern Echo.* http://www.thenorthernecho.co.uk/history/railway/stockton/316 5797.Efforts_that_kept_the_mines_afloat/

Farey, J. (1827). *A treatise on the steam engine: Historical, practical, and descriptive* Vol. 2.   Retrieved from https://archive.org/details/treatiseonsteame01fareuoft

Fleming, D. (1952). Latent Heat and the Invention of the Watt Engine. *Isis, 43*(1), 3-5. doi: 10.2307/227128

Fox, C. (2007). The Ingenious Mr Dummer: Rationalizing the Royal Navy in Late Seventeenth-Century England. *Electronic British Library Journal, Article 10*, 1-58. http://www.bl.uk/eblj/2007articles/pdf/ebljarticle102007.pdf

George, A. (2013). Penydarren Locomotive & Richard Trevithick. Retrieved October 5th, 2013, from http://www.alangeorge.co.uk/PenydarrenLocomotive.htm

Hatchett, C. (1967). *The Hatchett diary: a tour through the counties of England and Scotland in 1796 visiting their mines and manufactories*: D. Bradford Barton Ltd.

Howard, S. W. (2009). Jabez Carter Hornblower, Engineer and Inventor. Retrieved from Penrosefam.org website: http://penwood.famroots.org/jabez-carter-hornblower-7-27-09.pdf

Iltis, C. (1971). Leibniz and the Vis Viva Controversy. *Isis, 62*(1), 21-35. doi: 10.2307/228997

Kanefsky, J., & Robey, J. (1980). Steam Engines in 18th-Century Britain: A Quantitative Assessment. *Technology and Culture, 21*(2), 161-186. doi: 10.2307/3103337

Kelly, M. (2002). The non rotative beam engine. http://himedo.net/TheHopkinThomasProject/TimeLine/Wales/Steam/JamesWatt/Kelly/Watt.htm

Kirby, M. W. (2002). *The Origins of Railway Enterprise: The Stockton and Darlington Railway 1821-1863*: Cambridge University Press.

Kitsikopoulos, H. (2013). From Hero to Newcomen: The Critical Scientific and Technological Developments That Led to the Invention of the Steam Engine. *Proceedings of the American Philosophical Society, 157*(3),

304.

Kuhn, T. S. (1970). The structure of scientific revolutions. *Chicago and London*.

Lewis, M. (2006). Reflections on 1604. *Early railways, 3*, 8-22.

Merton, R. K. (1938). Science, technology and society in seventeenth century England. *Osiris, 4*, 360-632.

Nuvolari, A., & Tartari, V. (2011). *The Quality of English patents, 1617-1841: A Reappraisal using multiple indicators*. Paper presented at the Paper presented at the DIME Final conference.

Nuvolari, A., Verspagen, B., & von Tunzelmann, N. (2003). *The diffusion of the steam engine in eighteenth-century Britain*. Paper presented at the Applied evolutionary economics and the knowledge-based economy, Maastricht.

Oldroyd, D. R. (2007). *Estates, enterprise and investment at the dawn of the industrial revolution: estate management and accounting in the North-East of England, c. 1700-1780*: Ashgate Publishing, Ltd.

Papin, D. (1695). *Recueil de diverses pièces touchant quelques nouvelles machines* Retrieved from http://books.google.fr/

Ranger, T. O. (1957). Richard Boyle and the making of an Irish fortune, 1588-1614. *Irish Historical Studies, 10*(39), 257-297.

Rankine, J., & Rankine, W. (1862). *Biography of William Symington, civil engineer: inventor of steam locomotion by sea and land. Also, a brief history of steam navigation*: A. Johnston.

Rumford, B. C. o. R. (1798). An Inquiry concerning the Source of the Heat Which is Excited by Friction. . *Royal Society of London Philosophical Transactions Series I, 88*, 80-102.

Savery, T. (1827). The Miner's Friend: Or, an Engine to Raise Water by Fire. from http://archive.org/details/minersfriendora00savegoog

Scherer, F. M. (1965). Invention and Innovation in the Watt-Boulton Steam-Engine Venture. *Technology and Culture, 6*(2), 165-187. doi: 10.2307/3101072

Siegfried, R. (1988). The Chemical Revolution in the History of Chemistry. *Osiris, 4*, 34-50. doi: 10.2307/301742

Smiles, S. (1865). *Lives of Boulton and Watt: Principally from the original Soho mss. Comprising also a history of the invention and introduction of the steam engine* Retrieved from http://ia700201.us.archive.org/11/items/livesofboultonwa00smilrich/livesofboultonwa00smilrich.pdf

Sweet, R. (1999). *The English Town: 1680-1840*: Longman.

Thomas, R. H. (1980). *The Liverpool & Manchester Railway*: BT Batsford.

Thurston, R. H. (1878). *A History of the Growth of the Steam-Engine* Vol. 24. Retrieved from http://himedo.net/TheHopkinThomasProject/TimeLine/Wales/

Steam/URochesterCollection/Thurston/index.html
Trevithick, F. (1872). *Life of Richard Trevithick: with an account of his inventions* (Vol. 1): E. & FN Spon.
Turvey, R. (2005). Horse traction in Victorian London. *The Journal of Transport History, 26*(2), 38-59.
Valenti, P. (1979). Leibniz, Papin, and The Steam Engine. http://www.schillerinstitute.org/educ/pedagogy/steam_engine.html
Weeks, L. H. (1904). *Automobile Biographies* Retrieved from http://www.gutenberg.org/files/41891/41891-h/41891-h.htm
Worcester, E. S., & Partington, C. F. (1825a). *The century of inventions of the Marquis of Worcester*: J. Murray.
Worcester, E. S., & Partington, C. F. (1825b). *The century of inventions of the Marquis of Worcester: From the original ms. with historical and explanatory notes and a biographical memoir*: J. Murray.

About the Author

In 1975 Bouke J. G. van der Kooij (1947) obtained his MBA (thesis: Innovation in SMEs) at the Interfaculteit Bedrijfkunde (nowadays part of the Rotterdam Erasmus University). In 1977 he obtained his MSEE (Thesis: Micro-electronics) at the Delft University of Technology.

He started his career as assistant to the board of directors of Holec NV, a manufacturer of electrical power systems employing about eight thousand people at that time. His responsibilities were in the field of corporate strategy and innovation of Holec's electronic activities. Travelling extensively to Japan and California, he became well known as a Dutch guru on the topic of innovation and microelectronics.

From 1982 to 1986, he was a member of the Dutch Parliament (Tweede Kamer der Staten Generaal) and spokesman on the fields of economic, industrial, science, innovation, and aviation policy. He became known as the first member introducing the personal computer in Parliament, but his work on topics such as the TNO Act, Patent Act, Chips Act, and others went largely unnoticed. After the 1986 elections and the massive loss for his party (VVD), he was dismissed from politics and became a part-time professor (Buitengewoon Hoogleraar) at the Eindhoven University of Technology. His field was the management of innovation. He also started his own company in 1986, Ashmore Software BV, developer of software for professional tax applications on personal computers.

After closing these activities in 2003, he became a real estate project developer and a real estate consultant in 2009 till his retirement in 2013. Innovation being the focus of attention during all his corporate, entrepreneurial, political, and scientific life, he wrote three books on the subject and published several articles. In his first book, he explored the technological dimension of innovation (the pervasive role of microelectronics). His second book focused on the management of innovation and the human role in the innovation process. And in his third book, he formulated "Laws of Innovation" based on the Dutch societal environment in the 1980s.

In 2012 he started studying the topic of innovation again. In 2013 he was accepted at the TU-Delft by Prof. Dr. Cees van Beers as a PhD candidate. His focus is on the theory of innovation, and his aim is to develop a multidimensional model explaining innovation. For this he creates extensive and detailed case studies observing the inventions of the steam engine, electromotive engines, information engines, and computing engines. He studies their characteristics from a multidisciplinary perspective (economic, technical, and social).

Van de Kooij is married and spends a great deal of his time working in the south of France.